The Jungles of Randomness

Also by Ivars Peterson

Fatal Defect: Chasing Killer Computer Bugs

Islands of Truth: A Mathematical Mystery Cruise

The Mathematical Tourist: Snapshots of Modern Mathematics

Newton's Clock: Chaos in the Solar System

The Jungles of Randomness

A Mathematical Safari

Ivars Peterson

John Wiley & Sons, Inc.
New York ▹ Chichester ▹ Weinheim ▹ Brisbane ▹ Singapore ▹ Toronto

This book is printed on acid-free paper. ♾

Published by John Wiley & Sons, Inc.
Published simultaneously in Canada.

Library of Congress Cataloging-in-Publication Data
Peterson, Ivars.
The jungles of randomness: a mathematical safari
Ivars Peterson.
p. cm.
Includes bibliographical references and index.
ISBN 0-471-29587-6 (alk. paper)
1. Probabilities—Popular works. I. Title.
QA273.15.P48 1997
519.2—dc21 97-1275

Printed in the United States of America

10 9 8 7 6 5 4 3 2 1

To my mother,
Zelma Peterson,
for her fascination with the way things work.

Contents

Preface
Infinite Possibility

All mimsy were the borogoves,
And the mome raths outgrabe.

—*Lewis Carroll (1832–1898)*, "Jabberwocky"

An ape sits hunched over a keyboard. A long hairy finger bangs a key, and the letter *a* appears on the computer screen. Another random stab produces *n*, then a space, then *a*, *p*, and *e*. That an ape would generate this particular sequence of characters is, of course, highly improbable. In the realm of random processes, however, any conceivable sequence of characters is possible, from utter gibberish to the full text of this book.

The seemingly infinite possibilities offered by randomness have long intrigued me. Years ago when I was a high school student, I came across a provocative statement by Arthur Stanley Eddington (1882–1944), a prominent astronomer and physicist. "If an army of monkeys were strumming on typewriters, they *might* write all the books in the British Museum," he noted.

Eddington wanted to emphasize the improbability of such an outcome, and his remark was meant as an example of something that could happen in principle but never in practice. I was left with the indelible image of a horde of monkeys forever pecking away at typewriters, generating the world's literature.

The preface that you are now reading contains nearly two thousand words, or roughly ten thousand characters, including spaces. An ape at a keyboard can choose among twenty-six letters and about fourteen other keys for punctuation, numbers, and spaces. Thus, it has one chance in forty of hitting *a* as the first letter, one chance in forty of picking *n* next, and so on. To write the entire preface, the ape would have to make the correct choice again and again. The probability of such an occurrence is one in forty multiplied by itself ten thousand times, or one in $40^{10,000}$. That figure is overwhelmingly larger than the estimated number of atoms in the universe, which is a mere 10^{80}.

One would have to wait an exceedingly long time before a member of a troop of apes happened to compose this book by chance, let alone the millions of volumes in the Library of Congress and the British Museum. Sifting through the troop's vast output to find the flawless gems, including original works of significant merit, would itself be a notably frustrating, unrewarding task. By eschewing randomness, a human author, on the other hand, can generate a meaningful string of characters far more efficiently than an ape, and the output generally requires considerably less editing.

Most people, including mathematicians and scientists, would say that they have a good idea of what the word *random* means. They can readily give all sorts of examples of random processes, from the flipping of a coin to the decay of a radioactive atomic nucleus. They can also list phenomena in which chance doesn't appear to play a role, from the motion of Earth around the sun to the ricochets of a ball between the cushions of a billiard table and the steady vibrations of a violin's plucked string.

Often, we use the word *random* loosely to describe something that is disordered, irregular, patternless, or unpredictable. We link it with chance, probability, luck, and coincidence. However, when we examine what we mean by *random* in various contexts, ambiguities and uncertainties inevitably arise. Tackling the subtleties of randomness allows us to go to the root of what we can understand of the universe we inhabit and helps us to define the limits of what we can know with certainty.

We think of flipping a coin as a way of making a blind choice, yet in the hands of a skilled magician the outcome may be perfectly predictable. Moreover, a process governed entirely by chance can lead to a completely ordered result, whether in the domain of monkeys pounding on keyboards or atoms locking into place to form a crystal. At the same time, a deterministic process can produce an unpredictable outcome, as seen in the waywardness of a ball rebounding within the confines of a stadium-shaped billiard table or heard in the screeches of an irregularly vibrating violin string. We can even invent mathematical formulas to generate predictable sequences of numbers that computers can then use to simulate the haphazard wanderings of perfume molecules drifting through the air.

It's useful to distinguish between a random process and the results of such a process. For example, we think of typing monkeys as generators of random strings of characters. If we know that such a random process is responsible for a given string, we may be justified in labeling or interpreting the string itself as random. However, if we don't know the source of a given string, we are forced to turn to other methods to determine what, if anything, the string means. Indeed, reading itself involves just such a search for meaning among the lines of characters printed on a page or displayed on a computer screen.

Consider the passage from Lewis Carroll's poem "Jabberwocky" that starts off the preface. From the presence of a few familiar words, the pattern of spaces, and the vowel-consonant structure of the remaining words, we would surmise that the author intended those lines to mean something, even though we don't understand many of the words. If the same passage were to come from typing monkeys, however, we might very well reject it as gibberish, despite the fragments of structure and pattern.

Similarly, in flipping a coin we know from experience (or theory) that we're likely to obtain an equal number of heads and tails in a long sequence of tosses. So if we see twenty-five heads in a row, it might be the legitimate though improbable result of a random process. However, it might also be advisable to check whether the coin is fair and to find out something about the fellow who's doing the flipping. The context determines how we interpret the data.

At the same time, just because we happen to see a sequence of roughly equal numbers of heads and tails doesn't mean that the results arise from tosses of a fair coin. It's possible to program a computer with a numerical recipe that involves no randomness yet gives the same distribution of heads and tails. Thus, given an arbitrary sequence of heads and tails, there's really no way to tell with confidence whether it's the result of a random process or it's been generated by a formula based on simple arithmetic.

"From a purely *operational* point of view . . . the concept of randomness is so elusive as to cease to be viable," the mathematician Mark Kac said in a 1983 essay on the nature of randomness. Kac also took a critical look at the different ways in which we sometimes interpret randomness in different contexts. For example, in the book *Chance and Necessity*, the biologist Jacques Monod (1910–1976) suggested that a distinction be made between "disciplined" chance, as used in physics to describe, say, radioactive decay, and "blind" chance. As an example of the latter, he cited the death of a doctor who, on his way to see a patient, was killed by a brick that fell from the roof of a building.

Kac argued that the distinction Monod makes really isn't meaningful. Although statistics on doctors killed by falling bricks aren't readily available, there are extensive data on Prussian soldiers kicked to death by horses—events that also fall under the category of blind chance. When one compares data on the number of soldiers killed in specified time intervals with data on the number of radioactive decays that have occurred in analogous periods, the two distributions of events look very similar.

Mathematics and statistics provide ways to sort through the various meanings of randomness and to distinguish between what we can and cannot know. They help us shape our expectations in different situations. In many cases, we find that there are no guarantees, only probabilities. We need to learn to recognize such limitations on certainty.

The Jungles of Randomness offers a random trek through the mélange of order and disorder that characterizes everyday experience. Along the way, my intention is to reveal a little of the immense, though often overlooked, impact of mathematics on our lives, to examine its power to explain, to suggest its austere elegance and beauty, and to provide a glimmer of its fundamental playfulness.

The search for pattern is a pervasive theme in mathematics. It is this pursuit that brings to our attention the curious interplay of order hidden in randomness and the randomness that is embedded in order. It's part of what makes mathematics such an alluring sport for mathematicians.

My aim is to provide a set of mathematical X rays that disclose the astonishing scope of randomness. The mathematical skeletons unveiled in these revealing snapshots serve as a framework for understanding a wide range of phenomena, from the vagaries of roulette wheels to the synchronization of cells in a beating heart. It's like opening up a watch to see what makes it tick. Instead of gears, levers, and wheels, however, we see equations and other pieces of mathematical apparatus.

Characterizing the vibrations of a drum's membrane, arranging points on the surface of a sphere, modeling the synchronized blink of a cloud of fireflies in Thailand, and playing games of chance are among the mathematical pastimes that provide connections to various aspects of everyday life. Each of those playful activities has prompted new thinking in mathematics. Each one brings randomness into play.

Mathematics encompasses the joy of solving puzzles, the exhilaration of subduing stubborn problems, the thrill of discerning patterns and making sense of apparent nonsense, and the immense satisfaction of nailing down an eternal truth. It is above all a human enterprise, one that is sometimes pursued simply for its own sake with nary a practical application in mind and sometimes inspired by a worldly concern but invariably pushed into untrodden territory. Mathematical research continually introduces new ideas and uncovers intriguing connections between old, well-established notions. Chance observations and informed guesses develop into entirely new fields of inquiry. Almost miraculously, links to the rest of the world inevitably follow.

With its system of theorems, proofs, and logical necessity, mathematics offers a kind of certainty. The tricky part lies in establishing meaningful connections between the abstract mathematical world that we create in our minds and the everyday world in which we live. When we find such links, mathematics can deliver accurate descriptions, yield workable solutions to real-world problems, and generate precise predictions. By making connections, we breathe life into the abstractions and symbols of the mathematicians' games.

Intriguingly, the mathematics of randomness, chaos, and order also furnishes what may be a vital escape from absolute certainty—an opportunity to exercise free will in a deterministic universe. Indeed, in the interplay of order and disorder that makes life interesting, we appear perpetually poised in a state of enticingly precarious perplexity. The universe is neither so crazy that we can't understand it at all nor so predictable that there's nothing left for us to discover.

So, the trek through the jungles of randomness starts with games of chance. It proceeds across the restless sea of life, from the ebb and flow of human concourse to the intricacies of biological structure and the dynamics of flashing fireflies. It wanders into the domain of sounds and oscillations and the realm of fractals and noise. It emerges from the territory of complete chaos as a random walk. Glimpses of gambling lead to a lifetime of chance.

Let the games begin!

1

The Die Is Cast

"Iacta alea est!"

—*Julius Caesar (100–44* B.C.*)*

A die tumbles out of a cupped hand, bounces on the carpet a few times, rolls a short distance, then teeters to a stop. The uppermost face of the white cube shows four black dots arranged in a square.

Grinning, a child briskly moves a red token four squares to the right along the bottom row of a large checkerboard grid. The token lands on a square marked with the foot of a ladder. The player immediately takes the shortcut, advancing the token up the rungs to a higher row. Just ahead lies a square ominously marked with the upper end of a chute—the start of a costly detour.

With moves governed entirely by the roll of a die, Chutes and Ladders is a racecourse on which children of different ages and their elders can meet on an equal footing. Physical prowess and breadth of knowledge are immaterial on such a field. Only luck comes into play.

The playing of games has a long history. One can imagine the earliest humans engaged in contests of physical strength and endurance, with children racing about playing tag and great heroes struggling against daunting obstacles, as recorded in ancient myths. Written references to games go back thousands of years, and archaeologists have recovered a wide variety of relics that they interpret as gaming boards and pieces.

In the year 1283, when the king of Castile, Alfonso X (1221–1284), compiled the first book of games in European literature, he testified to the importance of games-playing in medieval society. "God has intended men to enjoy themselves with many games," he declared in the book's introduction. Such entertainments "bring them comfort and dispel their boredom."

Even in Alfonso's time, many of the board games he described were already hundreds of years old. Chess, the king's personal favorite, had been developed in India centuries earlier. Backgammon, one of the great entertainments of thirteenth-century nobility, had evolved from the Roman game *tabula*.

Succeeding centuries brought new amusements, along with variations on old ones. Each age and place had its particular favorites: the dice-and-counter game of pachisi in India, the coin-sliding game of shove ha'penny in William Shakespeare's England, the ancient game of go in China and Japan, and the card game cribbage in seventeenth-century Europe and America. In the Victorian era in Great Britain, nearly every parlor featured a wooden board of holes and pegs for the game of peg solitaire.

Amusement remains the motivation underlying the explosion of ingenuity that has now created a bewildering array of addictive computer, video, and arcade games, various forms of online and casino gambling, and new sports ranging from beach volleyball to snowboarding, along with novel board games and puzzles to tickle the mind.

"With their simple and unequivocal rules, [games] are like so many islands of order in the vague untidy chaos of experience," the novelist Aldous Huxley (1894–1963) wrote a few decades ago. "When we play games, or even when we watch them being played by others, we pass from the incomprehensible universe of given reality into a neat little man-made world, where everything is clear, purposive and easy to understand."

In these miniature worlds, competition brings excitement. Randomness serves as an equalizer. Chance introduces an element of suspense. Risk amplifies the thrill of play to an intoxicating level.

These tidy microcosms also attract mathematicians, who can't resist the distinctively human pleasure of learning the secrets of games. Who stands to win? What's the best move? Is there an optimal strategy? How long is a game likely to take? How do rules combine with chance to produce various outcomes? How are fairness and randomness linked?

In games of chance, each roll of a die, toss of a coin, turn of a card, or spin of a wheel brings a delicious surprise. Anyone can play. Anyone can win—or lose. Mathematics helps dispel some of the mystery surrounding unpredictable outcomes. It embodies an ever-present urge to tame the unruliness of Lady Fortune.

Using mathematical reasoning, we can't predict the outcome of a single roll of a die, but we can alter our expectations in the light of such an analysis. We may take comfort in the notion that, if the die is fair, each face of it will come up equally often in the long run. More generally, we can begin to make sense of and exploit the patterns that inevitably appear and disappear among the infinite possibilities offered by random choices.

Rolls and Flips

Dice are among the oldest known randomizers used in games of chance. In 49 B.C., when Julius Caesar ordered his troops across the river Rubicon to wage civil war in Italy, the *alea* of the well-known proverb he quoted already had the standard form of the die we use today: a cube engraved or painted with one to six dots, arranged so that the number of dots on opposite faces totals seven and the faces marked with one, two, and three dots go counterclockwise around a corner.

More than two thousand years earlier, the nobility of the Sumerian city of Ur in the Middle East played with tetrahedral dice. Carefully crafted from ivory or lapis lazuli, each die was marked on two of its four corners, and players presumably counted how many marked or unmarked tips faced upward when these dice were tossed. Egyptian tombs have yielded four-sided pencils of ivory and bone, which could be flung down or rolled to see which side faces uppermost. Cubic dice were used for games and gambling in classical Greece and in Iron Age settlements of northern Europe.

Because it has only two sides, a coin is the simplest kind of die. Typically, the two faces of a coin are made to look different (heads or tails), and this distinguishing feature plays a key role in innumerable pastimes in which a random decision hinges on the outcome of a coin toss.

How random is a coin toss? Using the equations of basic physics, it's possible to predict how long it takes a coin to drop from a known height. Apart from a small deviation due to measurement error, the time it will hit can be worked out precisely. On the other hand, a properly flipped coin tossed sufficiently high spins so often during its flight that calculating whether it lands heads or tails is practically impossible, even though the whole process is governed by well-defined physical laws.

Despite this unpredictability for individual flips, however, the results of coin tossing aren't haphazard. For a large number of tosses, the proportion of heads is very close to ½.

In the eighteenth century, the French naturalist Georges-Louis Leclerc (1707–1788), the Comte de Buffon, tested this notion by experiment. He tossed a coin 4,040 times, obtaining 2,048 heads (a proportion of 0.5069). Around 1900, the English mathematician Karl Pearson (1857–1936) persevered in tossing a coin twenty-four thousand times to get 12,012 heads (0.5005). During World War II, an English mathematician held as a prisoner of war in Germany passed the time in the same way, counting 5,067 heads in ten thousand coin tosses.

Such data suggest that a well-tossed fair coin is a satisfactory randomizer for achieving an equal balance between two possible outcomes. However, this equity of outcome doesn't necessarily apply to a coin that moves along the ground after a toss. An uneven distribution of mass between the two sides of the coin and the nature of its edge can bias the outcome to favor, say, tails over heads. A U.S. penny spinning on a surface rather than in the air, for example, comes up heads only 30 percent of the time. To ensure an equitable result, it's probably wise to catch a coin before it lands on some surface and rolls, spins, or bounces to a stop.

Empirical results from coin-tossing experiments support the logical assumption that each possible outcome of a coin toss has a probability of $\frac{1}{2}$, or .5. Once we make this assumption, we can build abstract models that capture the probabilistic behavior of tossed coins—both the randomness of the individual tosses and the special kind of order that emerges from this process.

Consider what happens when a single coin is tossed repeatedly. On the first toss, the outcome is either a head or a tail. Two tosses have four (2×2) possible outcomes, each with a probability of $\frac{1}{4}$ (or .25); and three tosses have eight ($2 \times 2 \times 2$) possible outcomes. In general, the number of possible outcomes can be found by multiplying together as many 2s as there are tosses.

One can readily investigate the likelihood that certain patterns will appear in large numbers of consecutive tosses. For example, if a coin is tossed, say, 250 times, what's the longest run of consecutive heads that's likely to arise?

A simple argument gives us a rough estimate. Except on the first toss, a run of heads can begin only after a toss showing tails. Thus, because a tail is likely to come up about 125 times in 250 tosses, there are 125 opportunities to start a string of heads. For about half of these tails, the next toss will be a head. This gives us around sixty-three potential head runs. Roughly half the time, the first head will be followed by a second one. So, around thirty-two runs will consist of two heads or more. About half of these will contain at least one additional head, meaning that we will probably get sixteen runs of three heads or more, eight runs of at least four heads, four runs of at least five heads, two runs of six heads or more, and one run of seven heads or more.

That's actually a surprising result. People who are asked to write down a string of heads or tails that looks random rarely include sequences of more than four or five heads (or tails) in a row. In fact, it's generally quite easy to distinguish a human-generated sequence from a truly random sequence because the one that is written down by a human typically incorporates an insufficient number of long runs.

Possible Outcomes of Tossing a Coin One, Two, or Three Times

	Number of Heads	Probability
One Toss		
T	0	$\frac{1}{2}$
H	1	$\frac{1}{2}$
Two Tosses		
TT	1	$\frac{1}{4}$
TH	1	$\frac{1}{4}$
HT	1	$\frac{1}{4}$
HH	2	$\frac{1}{4}$
Three Tosses		
TTT	0	$\frac{1}{8}$
TTH	1	$\frac{1}{8}$
THT	1	$\frac{1}{8}$
HTT	1	$\frac{1}{8}$
THH	2	$\frac{1}{8}$
HTH	2	$\frac{1}{8}$
HHT	2	$\frac{1}{8}$
HHH	3	$\frac{1}{8}$

In tossing a fair coin, the probability of each outcome—head (H) or tail (T)—is equal. If we toss once, there are only two possible outcomes, each of which has a probability of $\frac{1}{2}$. Tossing twice, we have four possible outcomes, each having a probability of $\frac{1}{4}$. Tossing three times, we have eight possible outcomes, each having a probability of $\frac{1}{8}$. From the table, you can see that with three tosses, the probability of obtaining no heads is $\frac{1}{8}$, one head is $\frac{3}{8}$, two heads is $\frac{3}{8}$, and three heads is $\frac{1}{8}$.

So, although an honest coin tends to come up heads about half the time, there's a good chance it will fall heads every time in a sequence of two, three, or four tosses. The chances of that happening ten times in a row are much smaller, but it can still happen. That's what makes it tricky to decide, just from a record of the outcomes of a short sequence of tosses, whether such as string is a chance occurrence or it represents evidence that the coin is biased to always come up heads.

Random Fairness

Like coins, cubic dice are subject to physical laws. An unscrupulous player can take advantage of this physics to manipulate chance. A cheat, for instance, can control a throw by spinning a die so that a particular face remains uppermost or by rolling it so that two faces stay vertical. In each case, the maneuver reduces the number of possible outcomes. A grossly oversized die, in particular, is quite vulnerable to such manipulation. The standardized size of dice used in casinos may very well represent a compromise configuration—based on long experience—that maximizes the opportunity for fairness. Casinos and gambling regulations specify the ideal dimensions and weight of dice.

A cheat can also doctor a die to increase the probability of or even guarantee certain outcomes. References to "loaded dice," in which one side is weighted so that a particular face falls uppermost, have been found in the literature of ancient Greece. Nowadays casino dice are transparent to reduce the chances of such a bias being introduced.

Even without deliberately creating a bias, it's difficult to manufacture dice accurately without introducing some asymmetry or nonuniformity. Manufacturers of casino dice take great pains to assure quality. Typically 0.75 inch wide, a die is precisely sawed from a rectangular rod of cellulose or some other transparent plastic. Pits are drilled about 0.017 inch deep into the faces of the cube, and the recesses are then filled in with paint of the same weight as the plastic that has been drilled out. The edges are generally sharp and square.

In contrast, ordinary store-bought dice, like those used in children's games, generally have recessed spots and distinctly rounded edges. Because much less care goes into the fabrication of such dice, they are probably somewhat biased. Achieving fairness is even more difficult with polyhedral dice that have eight, twelve, or twenty faces, each of which must be manufactured and finished to perfection.

In principle, an unbiased cubic die produces six possible outcomes. It makes sense to use a mathematical model in which each face has an equal probability of showing up. One can then calculate other probabilities, including how often a certain number is likely to come up. Several decades ago, the Harvard statistician Frederick Mosteller had an opportunity to test the model against the behavior of real dice tossed by a real person. A man named Willard H. Longcor, who had an obsession with throwing dice, came to him with an amazing offer to record the results of millions of tosses. Mosteller accepted, and some time later he received a large crate of big manila envelopes, each of which contained the results of twenty thousand tosses with a single die and a written summary showing how many runs of different kinds had occurred.

"The only way to check the work was by checking the runs and then comparing the results with theory," Mosteller recalls. "It turned out [Longcor] was very accurate." Indeed, the results even highlighted some errors in the then-standard theory of the distribution of runs.

Because the data had been collected using both casino dice from Las Vegas and ordinary store-bought dice, it was possible to compare their performance not only with theory but also with each other and with a computer that simulated dice tossing. As it turned out, the computer proved to have a poor random-number generator (see chapter 9), whereas the Las Vegas dice were very close to perfect in comparison with theory.

A mathematical model allows us to analyze the behavior of dice in both the short term and the long run and to study how the randomness of tumbled dice interacts with the rules of various games to favor certain strategies and results.

In some versions of Chutes and Ladders (or its Snakes and Ladders counterpart), each player must roll a six in order to start the game and then roll again to make his or her first move. It may actually take a while for this particular number to materialize, but more often than not each player will succeed in four or fewer throws. The mathematics of chance reveals why. On each of four rolls of a single die, the probability that six will not come up is $5/6$. Because each roll of a die is independent of any other roll, the probability that six will not come up in four rolls is $5/6 \times 5/6 \times 5/6 \times 5/6$, or $625/1296$. Hence, the chance of getting a six is $1 - 625/1296 = 671/1296 = .5177$, a probability of more than 50 percent.

During the seventeenth century, a favorite gambling game in the salons of Paris involved betting with even odds that a player would throw at least one six in four rolls of a single die. The calculated probabilities demonstrate that anyone who made the bet could win handsomely in the long run, especially if the unsuspecting victims believed intuitively that it would take six rolls to get the desired result.

One variation of the game involved rolling a pair of dice to obtain a double six in twenty-four throws. In this case, however, the gambler would lose in the long run if he offered even money on the possibility. One of the players who were puzzled by such an outcome was Antoine Gombaud (1610–1685), the Chevalier de Méré. An old gamblers' rule said that two dice can come up in six times as many ways as one die can. Thus, if four is the critical number of throws in a game with one die to reach favorable odds, then six times four, or twenty-four, should be the critical number of throws in a game with two dice.

The chevalier suspected that twenty-four wasn't the right answer, and he worked out an alternative solution of the problem, looking at the thirty-six different throws possible with two dice. He wasn't entirely sure of his conclusion, however, so he asked his acquaintance Blaise Pascal (1623–1662), a mathematician and a philosopher, to check the reasoning.

Pascal determined that the probability of rolling a double six after twenty-four throws turns out to be less than $\frac{1}{2}$. No double six comes up in thirty-five out of the thirty-six possible outcomes for two dice. Throwing the dice twenty-four times means that the probability of not getting a double six is $\frac{35}{36}$ multiplied by itself twenty-four times, or

Possible Results of Throws with Two Dice

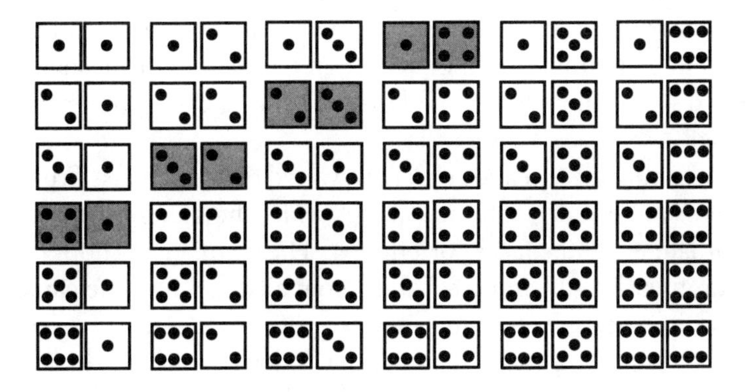

Rolling a fair die has six possible outcomes. In rolling two dice, for any given result with the first die, there are six possible outcomes for the second die, which gives a total of thirty-six pairs. What's the probability of rolling two dice to obtain, say, five? Because this total comes up in four different pairs (shaded), the probability is $\frac{4}{36}$.

$(35/36)^{24} = .5086$. So, the probability of winning is only $1 - .5086$, or .4914. Clearly, betting with even odds on such an outcome becomes a losing proposition in the long run.

Even with one die, however, it often takes a player many more than four throws to roll a particular number. On one occasion in five, a player will not succeed within eight throws, on one occasion in ten within twelve throws, and on one occasion in a hundred within twenty-five throws of the die.

Such random fluctuations in outcome can cause considerable anguish for young children in games in which a particular number must turn up at some point. A parent may find it prudent to bend the rules a little to forestall the tantrum or shower of tears that can result from pinning a child's hopes on random possibilities.

By the Rules

The origins of Chutes and Ladders go back many centuries to India and a board game called *moksha-patamu* (a Hindu concept that is akin to heaven and hell). Designed as a way of instructing children in religious values, the game graphically depicts the coexistence of good and evil and illustrates the impact of chance on human affairs.

In the game, checkerboard pilgrims follow a winding path that is shortened by virtuous deeds and lengthened by sinful acts or wrongdoing. They triumph when they climb ladders that represent faith, reliability, and generosity, and they fail when they slither down snakes that signify disobedience, vanity, and greed. A die dictates the fortunes of the road.

The game's design and rules have varied over the years. Typically a grid ten squares long and ten squares wide, the game board features several ladders and snakes (or chutes). The number of shortcuts and detours, their length, and their placement all affect how the game proceeds. A classic Indian design may have twelve snakes and eight ladders, with one long snake reaching back down to the starting square and a ladder of more modest length extending to the winning square at the top of the board.

In modern Western versions of the game, the moralizing usually takes on a gentler cast, and the snakes are often replaced by chutes. Usually the number of ladders equals the number of snakes, though in some upwardly mobile versions the ladders predominate. Whatever the design, most children see the game as an entertaining race spiced by forced detours.

In most of these game designs, the distribution and number of snakes and ladders probably reflects trial-and-error experience when it comes to the combination that makes a pleasing game. However, although rolls of a die determine the moves, it's possible to use mathematics to analyze the game and figure out where to place the snakes and ladders to ensure that the game is neither too long nor too short.

Of course, one could test a board design by playing hundreds of games and recording the results. Alternatively, one can use a mathematical theory named for the Russian mathematician Andrey A. Markov (1856–1922) to compute the average length of a game. Markov studied systems of objects that change from one state to another, according to specified probabilities. When an initial state can lead to subsequent states, and these states, in turn, can lead to additional states, the result is a Markov chain.

Suppose we examine a game of Chutes and Ladders in which there's only one player. Initially the player's token is off the board, and we call this condition state 0. The first roll of a die produces either one, two, three, four, five, or six, meaning that the player can move his token to one of the squares numbered 1, 2, 3, 4, 5, or 6 after the first roll.

If the player is lucky and rolls a one, he goes to square 1, and then up the ladder to square 38 without an additional turn. Rolling a four would again put him on a ladder, taking him to square 14. So, as the result of one turn the player can end up in state 2, 3, 5, 6, 14, or 38, with an equal probability of $\frac{1}{6}$. Squares 1 and 4 aren't considered states because the token doesn't come to rest there. Similarly, the bottom of any other ladder or the top of any chute doesn't count as a state.

For the second roll of the die, the game can be in one of six states, and we must consider each possibility separately. Suppose, for example, that the first throw puts the player in state 2. From there, the second roll moves him to square 3, 4, 5, 6, 7, or 8. Because 4 is the bottom of a ladder, the player again immediately ascends to 14. So, starting from state 2, the player can move to state 3, 5, 6, 7, 8, or 14 with equal probability.

If the first roll happened to put the player in state 14, the second roll would bring him to square 15, 16, 17, 18, 19, or 20 with equal probability. However, landing on square 16 would plunge him back to 6. Hence, from state 14 the player can move with equal probability to state 6, 15, 17, 18, 19, or 20.

After two rolls, the game can be in a total of twenty-one states. These states, however, are not equally likely. For example, there's only one way to arrive at state 44 at the end of two rolls: by rolling a one and then a six. Therefore, the probability of ending up in state 44 after two throws is $\frac{1}{6} \times \frac{1}{6} = \frac{1}{36}$. In contrast, there are three ways of getting to state 31: by rolling three and six, by rolling five and four, and by rolling six and three. Thus, the probability of being in state 31 after two throws is $(\frac{1}{6} \times \frac{1}{6}) + (\frac{1}{6} \times \frac{1}{6}) + (\frac{1}{6} \times \frac{1}{6}) = \frac{3}{36}$.

Such calculations can be carried out for three, four, or more rolls to create a complicated Markov chain showing how the game progresses from one set of states to another. For any position of the token, a player can readily determine the probability of ending up on that square.

One can also use the same technique to analyze particular situations. For instance, to place a chute so as to minimize a player's chance of climbing a particular ladder, it's possible to construct a table giving the probabilities of climbing a ladder when a ladder is a given number of squares ahead, first with no chutes in the way and then with a single chute on one of the intervening squares in front of the ladder.

A player who is one square in front of the ladder has one chance in six of rolling a one to get to the ladder, giving a probability of $\frac{1}{6}$, or about .17. A player two squares in front can get to the ladder by rolling either a one (probability of $\frac{1}{6}$) or two ones ($\frac{1}{36}$), for a combined probability of $\frac{1}{6} + \frac{1}{36}$, or .19. A player who starts at square three must roll a three ($\frac{1}{6}$), two and one ($\frac{1}{36}$), one and two ($\frac{1}{36}$), or one, one, and one ($\frac{1}{216}$), for a total probability of .23, to reach the ladder. If there is no chute, the probability of reaching the ladder turns out to be highest when the player is six spaces away.

One might think that placing a single chute immediately in front of a ladder would serve as a strong guard, but calculating the probabilities of landing there instead of on the ladder shows it to be the least effective position. A chute six squares in front is much more difficult to circumvent without missing the ladder as well.

Some of these factors come into play in Milton Bradley's Chutes and Ladders. The top row, or last ten squares, of the game has three chutes. One chute is two squares in front of the finish line. A second is five squares in front, and the third is eight in front. These chutes turn out to be effective guards, and more often than not a player wins by taking the ladder from square 80 directly to the winning hundredth square, completely bypassing the treacherous final stretch. Indeed, the likelihood of such a finish is increased by the fact that all three chutes in the top row land the player in the row that leads to the shortcut ladder in square 80, with no other obstacles standing in the way.

A bilingual version, called Gigantik Snakes & Ladders/Serpents et échelles and published by Canada Games, has a very different set of snakes and ladders. The game board has two places in which a guard snake immediately precedes a ladder, one near the beginning and another near the end. It also has two lengthy snakes slithering down from the top row and no ladder directly to the finishing square, so that games tend to take longer than those on the Milton Bradley board.

The theory of Markov chains offers a powerful method of analyzing probabilistic behavior in a wide variety of systems. Nowadays computers can readily handle large numbers of states, and researchers can quickly analyze the outcomes and fairness of such games as Chutes and Ladders and Monopoly. The same techniques can also be used in science to model the wanderings of molecules drifting in air, the foraging patterns of birds, and the fluctuation of prices on a stock market (see chapter 8).

Change of Face

The serious gamblers in casinos hang out at the craps tables. The basic rules of this two-dice game are simple, but the bewildering array of options for betting on various outcomes creates a fast-paced, insidiously seductive pastime in which a heady brew of chance, intuition, experience, calculation, and superstition come into play.

The shooter tosses two dice. If a total of seven or eleven comes up on a beginning roll, the shooter and those wagering with him win whatever amount they bet. If a two, three, or twelve total (called craps) shows up, the shooter and his companions lose. Players betting against the shooter win if a two or a three comes up. They neither lose nor win for a double six (twelve). Any of the remaining six totals (four, five, six, eight, nine, and ten) on a beginning roll starts off a different sequence of play, with different possible bets.

Suppose a shooter replaces the standard dice with a pair of new dice whose faces are marked as follows: 1, 2, 2, 3, 3, 4 and 1, 3, 4, 5, 6, 8. Should the other players or the casino management object?

First, we can check whether the probabilities of the various sums obtained with the new dice are different. We can do this by displaying in a table all the ways in which each sum from two to twelve can be obtained with the two different pairs of dice, as shown below. Determining the probabilities requires counting up the number of such totals on the chart and dividing by thirty-six.

Interestingly, the unusually marked dice and the standard dice have exactly the same frequencies for the possible sums. There's only one way to roll a two, two ways to roll a three, and so on.

Although the numbering on the weird dice is completely different from that on the standard dice, all the odds are exactly the same for rolling any given sum. You could make the substitution at a craps table as far as the sums are concerned. The trouble is that craps betting puts certain combinations, such as doubles, in special roles (for example, the double six on a beginning roll). With the weird dice, there's no way to get a double two, five, or six, and there are two ways to get a double three!

Changing the dice would have a considerable impact on many board games in which rolling doubles or other special combinations

Sums

Standard Dice

	1	2	3	4	5	6
1	2	3	4	5	6	7
2	3	4	5	6	7	8
3	4	5	6	7	8	9
4	5	6	7	8	9	10
5	6	7	8	9	10	11
6	7	8	9	10	11	12

Weird Dice

	1	2	2	3	3	4
1	2	3	3	4	4	5
3	4	5	5	6	6	7
4	5	6	6	7	7	8
5	6	7	7	8	8	9
6	7	8	8	9	9	10
8	9	10	10	11	11	12

Substituting a pair of "weird" dice with faces labeled 1, 2, 2, 3, 3, 4 and 1, 3, 4, 5, 6, 8 for a pair of standard dice has no effect on games in which only the sums matter. For both sets of dice, there's only one way to roll a sum of two, only two ways to roll a sum of three, and so on, as shown in charts of sums corresponding to the possible outcomes for standard dice (left) and weird dice (right). In games in which rolling doubles plays a role, however, there is a difference. There are six ways to roll doubles with standard dice and only four ways to roll doubles with weird dice (shaded squares).

affects a player's fate. In Monopoly, players buy, sell, rent, and trade real estate in a cutthroat competition to bankrupt their opponents. They take turns throwing a pair of dice, with the totals indicating how many spaces to proceed along an outside track that includes twenty-two properties, four railroads, two utilities, a Luxury Tax square, an Income Tax square, three Chance squares, and three Community Chest squares. Corner squares are marked Go, Just Visiting/In Jail, Free Parking, and Go to Jail.

Players start at Go. A double warrants a second throw, but three consecutive doubles sends a player directly to the In Jail square. To get out of jail, the player must throw a double. If he succeeds, whatever sum he gets decides how many spaces he can advance along the board.

However, using the nonstandard dice gives a lower probability of rolling doubles (only 4 out of 36 instead of 6 out of 36). Moreover, the chances of landing on a square six spaces away goes up twofold and the chances of landing four, ten, or twelve spaces away on a move out of jail are zero. Thus, if you happen to own the property St. James Place, which is six spaces away from jail, you are likely to collect lots of rent from players escaping jail. On the other hand, the owner of Virginia Avenue (four squares away from jail) loses out on this extra business.

In fact, according to calculations made more than a decade ago by two students of Joseph Gallian at the University of Minnesota–Duluth, the change in dice moves St. James Place up from the tenth to the sixth most frequently visited space in the game. Virginia Avenue descends from twenty-fourth to twenty-seventh in the rankings.

It's possible to prove mathematically that the weird dice represent the only alternative numbering of a pair that provides the same sum probabilities as standard dice. Of course, it doesn't matter how you arrange the six numbers on each die, though you can opt for symmetry by placing the numbers so that each set of opposite sides totals five or nine.

You can work out alternative numbering schemes not only for cubic dice but also for dice in the shape of a tetrahedron (four triangular faces), an octahedron (eight triangular faces), a dodecahedron (twelve pentagonal faces), and an icosahedron (twenty triangular faces). For example, a pair of standard octahedrons marked 1, 2, 3, 4, 5, 6, 7, and 8 have the same sum probabilities as a pair with one marked 1, 3, 5, 5, 7, 7, 9, 11 and the other marked 1, 2, 2, 3, 3, 4, 4, 5. Either set would work in a two-dice game.

There are also other ways to get the same sum results as two standard cubic dice. For example, you can use a tetrahedral die labeled 1, 1, 4, 4 combined with a spinner with 18 equal segments labeled 1, 2, 2, 3, 3, 3, 4, 4, 4, 5, 5, 5, 6, 6, 6, 7, 7, and 8. This combination yields the same sum probabilities as an ordinary pair of cubes labeled 1 through 6.

Using weird dice brings a fresh perspective to games of chance. It provides a welcome opportunity for examining the interaction between rules and randomizers.

The Long Run

Quite often, in a race game governed strictly by chance, the player who starts out ahead stays ahead for most, if not all, of the race. This striking feature is worth examining more closely.

In a two-player coin-flipping game, heads and tails will each win half the time, on the average. But in a game of a thousand flips, when the total number of heads versus the total number of tails is at issue, it's very likely that one player will be ahead in the cumulative scoring more than 90 percent of the time. In other words, the proportion of time that, say, the number of heads exceeds the number of tails can be very high. Although it's equally likely that either of the two players will be in the lead at any given moment, one of them will probably have the lead nearly the whole time.

Such a counterintuitive reality hinges on the fact that, like dice, cards, and roulette wheels, coins don't have memories. Each coin toss is independent of all other tosses. Even if the first few tosses happen to result in more heads than tails, the probability of getting either a head or a tail on each subsequent toss remains precisely ½. Whatever slight trend gets established at the beginning doesn't necessarily get washed out, even though, in the long run, the proportion of heads or tails generally gets closer and closer to ½ as the number of tosses increases.

Gamblers and other game players often reason that because the number of heads and tails even out over the long run, a sequence of, say, eight heads out of ten means that tails are somehow more likely to come up in the next ten or twenty turns to reestablish a proper balance. They imagine that the probabilities involved are somehow elastic, stretching more tightly or loosening as necessary to bring the results back in line with expected trends.

What people overlook is the fact that the proportion of heads and tails also approaches $\frac{1}{2}$ if, in subsequent tosses, heads and tails appear equally often—as they should. Suppose heads have shown up eight times and tails twice in the first ten flips. The proportion of heads is 0.8 at this stage. In the next two thousand turns, the players happen to toss a thousand heads (for a total of 1,008) and a thousand tails (a total of 1,002). Heads are still in the lead, but the proportion of heads is now considerably closer to $\frac{1}{2}$. Thus, flipping four heads in a row with a fair coin doesn't make it any more likely that the next flip will be tails. But it does make it more likely that you will end up with fifty-four heads in a hundred flips, as opposed to fifty.

Similarly, in a simple, two-player race game involving a single die, the player who's ahead at any given time has the higher probability of winning. Even at a hundred squares from the goal, a player merely four spaces behind faces odds of two to one against winning if it's his opponent's turn to throw the die.

Gamblers at roulette tables in casinos fall into the same kind of trap when they keep track of the numbers that happen to come up on the wheel. They usually believe they can improve their chances of guessing correctly if they can identify patterns in the sequences of numbers that come up. However, if the roulette wheel is truly unbiased, such efforts are fruitless. What happens on one turn of the wheel has no bearing on what occurs on the next.

Because the casino holds one or two numbers on the wheel as its own, the bank is certain to beat the majority of players in the long run. Those who continue to play long enough will be ruined sooner or later, and no system of betting and number or color selection can prevent such an outcome. Of course, apparent patterns do inevitably come up in the distribution of the numbers, but these ephemeral patterns don't mean a thing unless the wheel is imperfect (see Chapter 9).

The nineteenth-century Russian novelist Fyodor Dostoyevsky (1821–1881) became seriously hooked on roulette in 1863, when he won a considerable fortune on his first foray into playing the wheel. In a letter to his brother Misha, Dostoyevsky wrote:

> My dear Misha, in Wiesbaden I invented a system, used it in actual play, and immediately won 10,000 francs. The next morning I got excited, abandoned the system and immediately lost. In the evening I returned to the system, observed it strictly, and quickly and without difficulty won back 3,000 francs. . . .

> When I arrived in Baden-Baden I went to the tables and in *a quarter of an hour* won 600 francs. This goaded me on. Suddenly I began to lose, could no longer keep my head, and lost every farthing.

So it went in Dostoyevsky's erratic succession of roulette episodes across Europe—a journey marked by requests for loans, unpaid hotel bills, and pawned watches—until 1871, when he finally quit. His masterful short novel *The Gambler*, written in haste in 1865 to settle a debt, presents a striking portrait of an individual irresistibly drawn into Lady Fortune's web. Dostoyevsky describes a gambler's obsession with the zero slot on the roulette wheel:

> It was, of course, a rare happening for zero to come up three times out of some ten or so; but there was nothing particularly astonishing about it. I had myself seen zero turn up three times running two days before, and on that occasion one of the players, zealously recording all the coups on a piece of paper, had remarked aloud that no earlier than the previous day that same zero had come out exactly once in twenty-four hours.

Dostoyevsky was certainly not the first to write about the lure of the luck of the draw. Girolamo Cardano (1501–1576), a brilliant, highly regarded physician and mathematician with an impressive array of intellectual achievements to his credit, was another player who found gambling irresistible. Around 1520, Cardano wrote *The Book on Games of Chance*, in which he presented competent analyses of backgammon and several other dice games, along with a variety of card games that included an early version of poker. He also excelled at chess, which at that time was accompanied by much betting and handicapping.

In a section on who should play and when, Cardano offered the following advice about moderation in gambling:

> So, if a person be renowned for wisdom, or if he be dignified by a magistracy or any other civil honor or by a priesthood, it is all the worse for him to play. . . . You should play rarely and for short periods, in a suitable place, for small stakes, and on suitable occasions, or at a holiday banquet.

It was advice that Cardano himself had trouble heeding. Erratic, argumentative, and obsessed, he gambled incessantly, sending his family into the Milan poorhouse for long stretches. In his brutally frank *Autobiography*, he remarked:

> During many years—for more than forty years at the chess boards and twenty-five years of gambling—I have played not off and on but, as I am ashamed to say, every day. Thereby I have lost esteem, my worldly goods, and my time. There is no corner of refuge for my defense, except if someone wishes to speak for me, it should be said that I did not love the game but abhorred the circumstances which made me play: lies, injustices, and poverty, the insolence of some, the confusion in my life, the contempt, my sickly constitution and unmerited idleness, the latter caused by others.

The irrational superstitions and strategies of gamblers testify to the extraordinarily vast landscapes of possibility in games of chance. Unpredictability reigns, with patterns appearing out of nowhere, then mysteriously disappearing. Anything is possible, yet the possibilities are doled out in measured doses. Faced with the astonishing scope of chance, the compulsive gambler lives on the brink, ever focused on the next spin of the wheel or roll of the dice.

Using the mathematics of chance, one can predict outcomes, establish conditions for guaranteeing certain results, or, at the very least, show whether any particular game is worth playing.

Games provide structured situations in which people can compute probabilities, often with great precision, and test their hypotheses. They can then bring this knowledge to bear not only in games but also in everyday situations in which uncertainty and randomness play a role. Understanding the laws of probability can guide their decisions in a wide range of matters.

Of course, games teach nothing unless one pays attention to the odds and learns to deal with them. But even precise calculation may not be enough in the face of human greed and deceit. In his 1753 novel *The Adventure of Ferdinand Count Fathom,* Tobias Smollett (1721–1771) noted that his hero, Fathom, learned to "calculate all the chances with the utmost exactness and certainty," only to be fleeced by a team of swindlers. A knowledge of mathematics is no proof against human folly.

The theory of probability combines commonsense reasoning with calculation. It domesticates luck, making it subservient to reason. It stands as an attempt to master the vagaries of uncontrollable circumstance in the face of unavoidable ignorance. Nonetheless, the tantalizing patterns that come and go among the infinite possibilities of randomness have a way of entangling the human mind. Randomness and pattern intermingle in mysterious ways. Where there is one, the other inevitably lurks.

2

Sea of Life

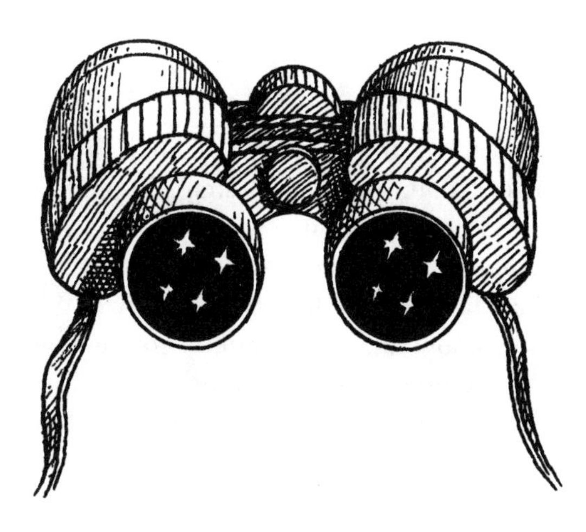

Like driftwood spars, which meet and pass
Upon the boundless ocean-plain,
So on the sea of life, alas!
Man meets man—meets and quits again.

—*Matthew Arnold (1822–1881)*, "Switzerland"

The tangled threads of human activity bring us together in many ways. We meet to share thoughts with family and friends. We gather to exchange information with colleagues or trade gossip with neighbors. We assemble to conduct business. We join in the camaraderie of team efforts. We jam into buses, trains, and planes. We flock together to drink in the crowd-spiked exhilaration of a great contest or an artful performance.

In work and play, from hour to hour and year to year, we pass from one gathering to another. Sometimes we are surrounded by people we know; sometimes we find ourselves in the presence of strangers. The restless ebb and flow of human concourse appears to be quite random, yet there are curious patterns present in the composition of these random groupings of strangers and acquaintances who come and go.

People Who Know People

Whether in large crowds, modest social gatherings, or small groups, people congregate in all sorts of combinations of strangers and acquaintances. Consider the varied possibilities in a party of six.

- ▷ Three couples stand in line to buy tickets to a movie. The members of each couple know each other, but no couple is acquainted with either of the other couples. Among the six people, each knows one other person but not the remaining four.
- ▷ A newcomer arrives at a club meeting. She introduces herself to each of the five members already present. They are all strangers to her, but the five veteran members each know the others quite well.

In each of the above examples, the sextet includes a group of at least three people, all of whom know one another, *or* a group of at least three people, none of whom know one another. Interestingly, *all* possible combinations of mutual strangers and acquaintances that can arise among six people obey the same rule: Any party of six must contain at least three acquaintances or at least three strangers.

To prove this, we could write down all possible combinations of strangers and acquaintances, starting with a group in which everyone knows one another (six acquaintances). However, there are 32,768 such possibilities. This number reflects the fact that any pair in a group of six must be either acquainted or strangers. Moreover, there are six choices of people for the first member of a pair, which leaves five choices for the second member of the pair, for a total of 30 possibilities. Because the order of the choices in a particular case doesn't matter (picking Alice first and then Barry is the same as picking Barry, then Alice), the number of possibilities that count is down to 30/2, or 15. Overall, for two possible relations (stranger or acquaintance), the number of possibilities is 2^{15}, or 32,768.

Luckily there's a shorter path to a proof that a group of six is sufficiently large to guarantee the presence of three mutual strangers or three mutual acquaintances. We need consider only two particular cases. Suppose a party is attended by the conveniently named guests Alice, Barry, Charles, Diana, Edith, and Frank. Of the five partygoers Barry, Charles, Diana, Edith, and Frank, either at least three are friends of Alice or at least three are strangers to Alice. Suppose Alice knows Barry, Charles, and Diana. If either Barry and Charles, Barry and Diana, or Charles and Diana are friends, then Alice and the acquainted pair make three people who know one another. Otherwise Barry, Charles, and Diana are mutual strangers.

Further, suppose Alice knows only two (or fewer) of the others, say, Barry and Charles. If either Diana and Edith, Diana and Frank, or Edith and Frank are strangers, then Alice and the unacquainted pair make three people who do not know one another. Otherwise Diana, Edith, and Frank are mutual acquaintances. This argument covers all possible cases for each of the six dinner-party guests. The rule holds every time: Any party of six people must include at least three mutual acquaintances or three mutual strangers.

The result represents a special case stemming from a branch of pure mathematics known as Ramsey theory. Instead of talking about six people, we can consider groups consisting of any number of people. Instead of specifying just two relationships (acquaintances and strangers), we can have any number of mutually exclusive relations. For example, we could consider people (or nations) who are allies, foes, and neutral parties—representing three mutually exclusive relationships—embroiled in a widespread conflict.

In general, Ramsey theory concerns the existence of highly regular patterns in sufficiently large sets of randomly selected objects, whether they are gatherings of people, piles of pebbles, stars in the night sky, or sequences of numbers generated by the throw of a die. In the party problem described above, for example, a group consisting of at least six members is needed in order to guarantee that there are three friends or three strangers present. No such guarantee is possible if the gathering has only five people.

Patterns can arise out of randomness in a variety of ways. On a clear, moonless night, we see thousands of stars scattered across the sky. With so many stars visible, it's not particularly difficult to pick out groups that appear in a certain pattern: four stars that very nearly form a straight line or a square, six stars that define a cross, seven stars in the shape of a dipper. The human imagination fills in the rest, connecting the dots to create a menagerie of celestial creatures that inhabit the sky.

We can achieve the same effect by randomly sprinkling dots across a sheet of paper. Given enough dots, we can find the same patterns on our sheet that we see in the night sky. Indeed, the human mind has a predilection for discerning patterns in the midst of disorder, whether in the shape of a humanlike face in a pile of rocks on the surface of Mars or a ghostly galleon of a cloud sailing the sky on a moonlit night.

Sequences of random numbers also inevitably display certain regularities. Consider, for example, the results of rolling a die or tossing a coin (see chapter 1). In sixty tosses of a coin, it's not unusual to have at least five tails or five heads in a row, or, in two thousand turns, to obtain a string of ten consecutive tails or heads.

Such patterns become even more striking when we plot the results of successive flips of a coin on a checkerboard grid, filling in the

Tabulating the outcomes of 2,500 flips of a coin by sequentially filling in the squares of a 50-by-50 grid (black for tails, white for heads) produces a random array. It's not difficult, however, to pick out fragments of patterns among the black or white squares, in much the same way that you can find patterns among stars randomly scattered across the sky. In this case, one array represents tosses of a fair coin and the other array represents tosses of a slightly biased coin. Can you tell which is which?

squares one at a time, row by row. By coloring the squares (say, black for tails and white for heads), one is highly likely to identify a variety of intriguing regularities among the clumps of like-colored squares scattered across the sheet.

Ramsey theory implies that complete disorder is impossible. Somehow, no matter how complicated, chaotic, or random something appears, deep within that morass lurks a smaller entity that has a definite structure. Striking regularities are bound to arise even in a universe that has no rules.

Slot Congestion

Sorting the mail that comes into an office generally requires that each piece be slipped into the appropriate slot of an array of pigeonholes—one for each employee. Suppose that a small business needs ten such slots. When eleven pieces of mail arrive, one or more of the slots will have to contain at least two items. So, if there are more pigeons than holes, some of the pigeons have to double up. Expressed mathematically, the so-called pigeonhole principle is a handy idea for proving theorems, and it often comes up in Ramsey theory to demonstrate the inevitability of the presence of certain patterns.

Perhaps the simplest possible application of the pigeonhole principle concerns groups of men and women. If there are three people present in a gathering, at least two must be men or at least two must be women. The same principle can be applied to any allocation of a given type of object, whether it is balls dropped into boxes, people slotted into certain categories, or numbers meeting particular criteria.

For a more subtle, nontrivial use of the pigeonhole principle, we can look at sequences of numbers. Consider, for example, the set of numbers 0, 1, 2, 3, 4, 5, 6, 7, 8, and 9. In this case, the numbers are written as an increasing sequence. Each successive number is larger than its predecessor. The numbers can also be written as a decreasing sequence: 9, 8, 7, 6, 5, 4, 3, 2, 1, 0.

Now, suppose the order of these numbers is randomly scrambled. Mathematicians argue that the resulting sequence of numbers will always contain a shorter sequence (or a subsequence) of at least four increasing or at least four decreasing numbers. For example, in the scrambled set 1, 2, 8, 0, 3, 6, 9, 4, 5, 7, the numbers 1, 2, 8, 9 represent an increasing subsequence. The scrambled sequence 8, 4, 9, 2, 1, 6, 3,

7, 0, 5 happens to contain a decreasing subsequence consisting of 8, 4, 2, 1, 0.

How can we be sure that an arbitrary sequence of ten different numbers will have an increasing or a decreasing subsequence consisting of at least four numbers? The number of differently ordered sets containing ten members is $10 \times 9 \times 8 \times 7 \times 6 \times 5 \times 4 \times 3 \times 2 \times 1$ (or 10!), which equals 3,628,800. As in the party-of-six example, the array of possible arrangements is too large for each one to be checked individually.

The most direct, elegant proof of this claim hinges on the pigeonhole principle. Consider the sequence 3, 0, 6, 2, 1, 8, 7, 9, 5, 4. The idea is to put the numbers in an array of slots so that consecutive numbers in each row increase going to the right and in each column decrease going down, as shown in the figure below. The first number goes into the slot in the upper-left-hand corner.

Increasing →

Decreasing ↓

3	6			
0	2			

The next number, 0, is smaller than 3, so it goes in the slot below 3. The following number, 6, is larger than 0, so it can't go in the slot beneath 0. It goes in the slot to the right of 3.

The number 2 is larger than 0 but smaller than 6, so it fits into the slot to the right of 0 and below 6. Similarly, 1 is smaller than 2 and 6 and larger than 0, so the only place it fits according to the allocation rule is in the slot below 2. Each number in turn can be slotted into the array, and from the result it's easy to read off both increasing (3, 6, 8, 9) and decreasing (8, 7, 5, 4) subsequences of length four, as shown in the figure on the next page.

Increasing →

Decreasing ↓

3	6	8	9	
0	2	7		
	1	5		
		4		

Because every number has to fit into its own slot, the diagram suggests that the very best that anyone can do is to cram nine of the numbers into a three-by-three corner, leaving the tenth number to complete either an increasing or a decreasing subsequence of length four. In general, there has to be at least one such subsequence in any scrambled set of ten consecutive numbers.

For strings of 101 arbitrarily ordered numbers, one finds increasing or decreasing subsequences of length eleven or more, and that's not necessarily true of a sequence of just a hundred different, scrambled numbers. In general, it's possible to prove that for $n^2 + 1$ numbers in a sequence, one will always get increasing or decreasing subsequences of length at least $n + 1$. In the example given above, $n = 3$, so sequences have $3^2 + 1$, or 10, members and subsequences have a length of at least $3 + 1$, or 4.

The pigeonhole principle is one tool that mathematicians can use to identify certain patterns that are inevitably present in a random array. In a more general sense, it also applies to everyday situations. Anyone who frequently plays solitaire card games has undoubtedly noticed that when the cards start falling into place early, the game more often than not proceeds rapidly to a successful conclusion. At a certain stage, victory becomes nearly certain as the number of cards available for play decreases rapidly and open slots get filled, leaving chance a smaller field on which to confound the ultimate result. In some games, there's no need to unveil the last few hidden cards because a successful outcome is already guaranteed.

So, patterns and predictability can arise in a variety of ways even in the realm of pure chance. Ramsey theory involves establishing the guarantees for such unexpectedly regular behavior among sequences of numbers, in sets of objects, in geometric arrays of dots and lines, and in mathematical logic.

Playing Fields of Logic

Ramsey theory owes its name to Frank Plumpton Ramsey, an English mathematician, philosopher, and economist. His father, Arthur J. Ramsey, was a professor of mathematics and the president of Magdalene College at Cambridge University. Frank Ramsey was born in 1903 and spent nearly his entire life in Cambridge. After he graduated in 1925 as Cambridge University's top math scholar, he remained at the university, quickly developing a reputation as an enthusiastic lecturer who could present ideas with great clarity. Applying his keen insight, he made significant, wide-ranging contributions to the study of theoretical economics and several other fields.

One of Ramsey's contemporaries, the philosopher George E. Moore (1873–1958), wrote that Ramsey "combined very exceptional brilliance with very great soundness of judgment in philosophy. He was an extraordinarily clear thinker: no one could avoid more easily than he the sort of confusions of thought to which even the best philosophers are liable, and he was capable of apprehending clearly, and observing consistently, the subtlest distinctions."

Ramsey was fascinated by questions of logic. He had been deeply influenced by Bertrand Russell (1872–1970), a philosopher and mathematician who believed that mathematics is synonymous with logic. Russell's ambitious goal was to prove that "all pure mathematics deals exclusively with concepts definable in terms of a very small number of fundamental logical concepts, and that all its propositions are deducible from a very small number of fundamental logical principles." These words appear in Russell's book *The Principles of Mathematics*, in which he set forth his landmark thesis.

Russell was keenly interested in how we come to know what we know—how we draw valid conclusions from evidence. For him, such questions required close study of logic puzzles and paradoxes to help expose potential cracks in structures of belief. "A logical theory may be tested by its capacity for dealing with puzzles, and it is a wholesome plan, in thinking about logic, to stock the mind with as many puzzles as possible, since these serve much the same purpose as is served by experiments in physical science," he argued.

Russell was not the only one who was concerned about logic and the foundations of mathematics. At the beginning of the twentieth century, the German mathematician David Hilbert (1862–1943) had argued that there had to be a clear-cut procedure for deciding whether a given logical proposition follows from a given set of axioms. An axiom is a statement that is assumed to be true without proof.

Ramsey set out to demonstrate that such a decision process exists for at least one particular type of mathematical problem. As one of the steps in his argument, he considered relationships among sets of whole numbers, and he proved that if the number of objects in a set is sufficiently large and each pair of objects has one of a number of relations, there is always a subset containing a certain number of objects where each pair has the same relation. In other words, there are patterns implicit in any sufficiently large set or structure. We've already seen Ramsey's theorem applied in cases in which the objects are people (in groups of six) and whole numbers (in increasing and decreasing sequences).

Though Ramsey was probably aware that his theorem could be applied to many different areas of mathematics, he focused on its application to mathematical logic. He showed that, given a set of axioms, certain true relationships must exist within that logical system. Ramsey read his paper containing the result, "On a Problem of Formal Logic," to the London Mathematical Society in 1928, and it was published two years later. He died in 1930, at the age of twenty-six, as a result of complications associated with abdominal surgery for a grave illness. Yet he left an astonishing legacy of achievement in several fields. Ironically, the mathematical work for which he is now best known attracted little attention in his day.

Moreover, subsequent developments showed that Ramsey's success in finding a decision procedure for a particular class of mathematical problems was not representative of the general case. Just a few years after Ramsey's death, Kurt Gödel (1906–1978), followed by Alan Turing (1912–1954) and others, proved that no such decision procedure was possible for any logical system of axioms and propositions sufficiently sophisticated to encompass the kinds of problems that mathematicians work on every day. Typically, it's possible to make true statements or propose theorems that one can't prove within the given system without adding additional axioms (see chapter 10).

Although Ramsey's theorem is accurately attributed to Frank Ramsey, its initial popularization stemmed from its application in geometry and the playful antics of a group of young mathematicians in Hungary in the 1930s.

Planes of Budapest

Nearly every Sunday during the winter of 1933 in Budapest, a small group of students would meet somewhere in the city at a park or cafe to discuss mathematics. The gathering typically included Paul Erdös (1913–1996), who was attending the University of Budapest, Gyuri (George) Szekeres, a recent chemical engineering graduate of the Technical University of Budapest, and another student, Esther Klein. The group enjoyed exchanging personal gossip, talking politics, and feeding their passion for mathematics. In effect, the students had their own university without walls, and they relished the chance to explore new ideas.

On one particular occasion, Klein challenged the group to solve a curious problem in plane geometry that she had recently encountered. Suppose there are five points positioned anywhere on a flat surface, as long as no three of the points form a straight line.

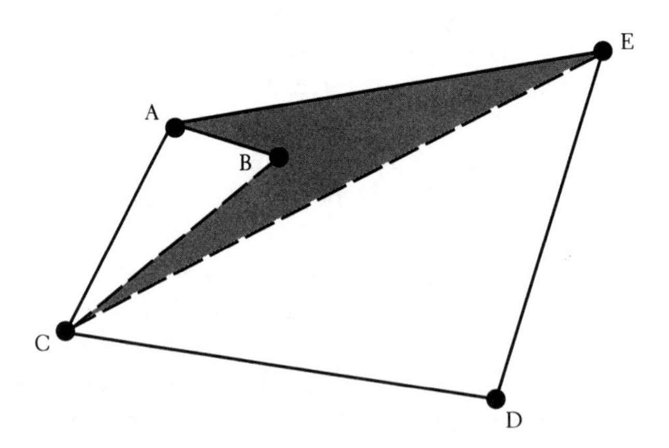

Given five points positioned on a flat surface so that no three points lie in a straight line, four of the points always define a convex quadrilateral, meaning that the resulting figure has no indentations. Here, ACDE is a convex quadrilateral, but ABCE (shaded) is not because of the indentation at B.

When four points are joined, they produce a quadrilateral—a four-sided figure. Klein had noticed that given five points, four of them always appeared to define a convex quadrilateral. The term *convex* means that the figure has no indentations. Thus, a square is convex, but a crescent quadrilateral (an angular boomerang) is not. In the example shown above, ACDE is a convex quadrilateral, but ABCE is not because of the indentation at B.

Klein asked if anyone could prove that a convex quadrilateral has to be present in any given set of five points lying in a plane, with no three points in a straight line. After letting her friends ponder the problem for a while, Klein submitted her proof.

She reasoned that there are three possible ways in which a convex polygon could enclose all five points, as is shown in the figure below. The simplest case occurs when this convex polygon results from joining four points to create a quadrilateral, which fences in the remaining point and automatically satisfies the requirement. In the second case, if the convex polygon is a pentagon incorporating all five points, then any four of these points can be connected to form a quadrilateral. Finally, if the convex polygon includes just three points to create a triangle, the two points left inside the triangle define a line that

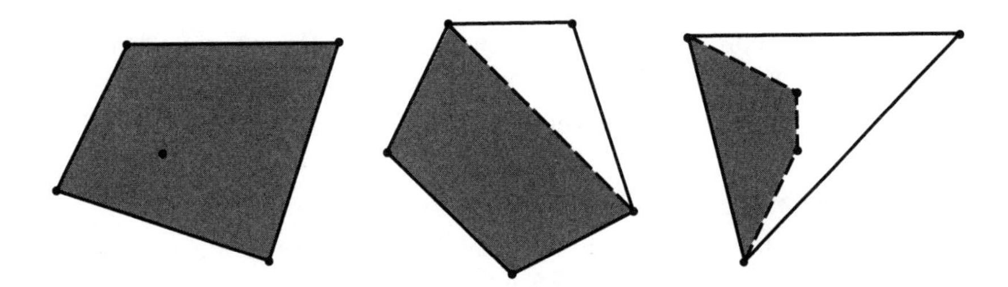

In proving that four of five points drawn on a flat surface always form a convex quadrilateral, imagine the points as nails sticking out of a board with an elastic band stretched so that it rings the largest possible number of nails. In the simplest case, the band touches four nails to create a quadrilateral, which fences in the remaining nail and satisfies the requirement (left). In the second case, the elastic band touches all five nails (middle). Here, connecting any four of the points forms a quadrilateral. Finally, the band can touch three nails, leaving the other two nails on the inside (right). In this case, the two points inside the triangle define a line that splits the triangle so that two of the triangle's points are on one side. These two points plus the two interior points automatically form a convex quadrilateral.

splits the triangle so that two of the triangle's points are on one side. These two points plus the two interior points automatically form a convex quadrilateral.

Impressed by Klein's ingenuity, Erdös and Szekeres quickly discovered a way of generalizing the result. They saw, for example, that five of nine points in a plane will always form a convex pentagon. Szekeres ended up demonstrating that no matter how randomly points may be scattered across a flat surface, one can always find a set that forms a particular polygon—if there are enough points. Thus, one can always find a convex pentagon in an array of nine points but not among eight. Erdös then proposed that if the number of points that lie in a plane is $1 + 2^{k-2}$, one can always select k points so that they form a convex polygon with k sides. Thus, for a quadrilateral, k is 4, and the number of points required will be $1 + 2^{4-2} = 1 + 2^2 = 5$.

In describing this occasion in a memoir written many years later, Szekeres recalled: "We soon realized that a simple-minded argument would not do, and there was a feeling of excitement that a new type of geometrical problem [had] emerged from our circle."

Szekeres had rediscovered Ramsey's theorem, though no one in the group realized it at the time. Instead of whole numbers or statements related to a system of axioms, the mathematical objects in this case are geometric figures consisting of points and lines and the substructures are specific convex polygons. Erdös dubbed the result the Happy End Theorem. The ending he had in mind was the marriage of Szekeres and Klein, who now live in Sydney, Australia.

Interestingly, no one has yet proved Erdös's conjecture concerning the precise number of points required to guarantee the presence of a given convex polygon. Erdös himself went on to become the most prolific and well-traveled mathematician of the twentieth century. He played a central role in the transformation of Ramsey theory, which involves the application of Ramsey's theorem and related propositions, from a collection of isolated results into a coherent body of work.

Though pursued mainly for the thought-provoking mathematical puzzles that it suggests, Ramsey theory has also begun to play a role in the design of vast computer and telephone networks. Techniques used in the theory can help engineers and computer scientists to take advantage of patterns that inevitably arise in large systems in which data storage, retrieval, and transmission have a significant random component.

Party Puzzles

Throughout his long, itinerant life, Paul Erdös spent most of his waking hours and, apparently, all of his sleeping hours doing mathematics. He was a superb problem solver, and his phenomenal memory allowed him to cite exact references to thousands of mathematical papers, including their page numbers. "If you don't know how to attack a particular problem, ask Erdös" was the constant refrain among his colleagues and collaborators. He was a one-man clearinghouse for information on the status of unsolved problems in the fields of mathematics that attracted his attention.

By the time he died in 1996, Erdös had also earned a reputation as the greatest problem poser of all time. He had the knack of asking just the right mathematical question—one that was neither so simple that it could be solved in a minute nor so difficult that a solution was unattainable. Ramsey theory, which offered a wide variety of challenges, was one of his joys.

A typical question involving Ramsey theory concerns how many numbers, points, or objects are needed to guarantee the presence of a certain structure or pattern. One looks for the smallest "universe" that automatically contains a certain object. For example, it may be the smallest group of arbitrarily positioned stars that would include a convex quadrilateral or the number of people required to ensure the attendance of three strangers or three acquaintances.

Fascinated by such problems, Paul Erdös particularly liked those he called "party puzzles," which involve different combinations of strangers and acquaintances brought together at some sort of gathering. But he didn't usually think of these problems in terms of people—except as a way of describing them informally. He reasoned in terms of points, lines, and colors—all elements of a branch of mathematics known as graph theory. In general, a graph consists of an array of points, or vertices, connected by straight lines, which are often called edges.

It's straightforward to convert a party puzzle into a graph problem. For instance, in the situation described at the beginning of this chapter, one can draw a point for each of the six people present in the group. These points can then be joined by lines, with each line colored red to signify two people who know each other or blue to mean the two people are strangers. When all pairs of points are

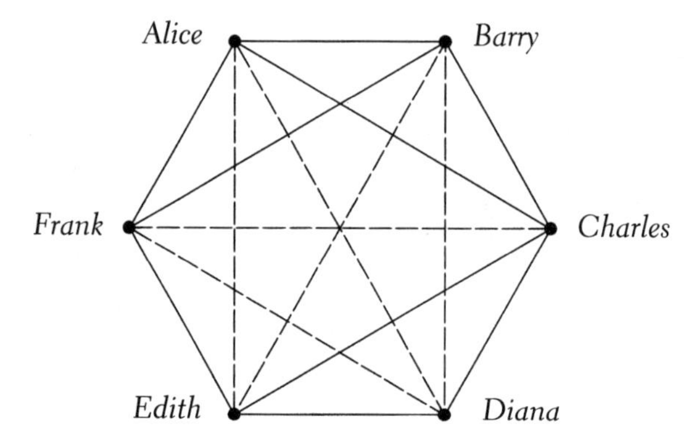

To show whether the guests at a party are mutual strangers or acquaintances, you can draw a graph in which each point stands for a person and the lines linking the points represent the relationship (solid for acquaintance or dashed for stranger). Depending upon the relationships specified in a given case, such a graph may contain just solid lines, just dashed lines, or a mixture of solid and dashed lines joining the points. This graph depicts relationships among six people at a party. The same graph can be drawn using different colors instead of solid and dashed lines.

joined, the resulting network of points and lines is known as a complete graph.

Depending on the relationships specified in a given case, such a graph may contain only red lines, only blue lines, or a mixture of red and blue lines joining the points. The problem comes down to proving that no matter how the lines are colored, one can't avoid producing either a red triangle (representing three mutual acquaintances) or a blue triangle (three strangers).

Such a coloring is guaranteed for a complete graph of six points, but not necessarily for a graph of five or fewer points, as the following figure demonstrates. Hence, the minimum number of points that have the requisite pattern is six, often designated by the so-called Ramsey number R(3,3), where the first number inside the parentheses gives the number of acquaintances and the second gives the number of strangers.

What about subgroups of four strangers or four acquaintances? It turns out that an array of seventeen points is insufficient to guarantee

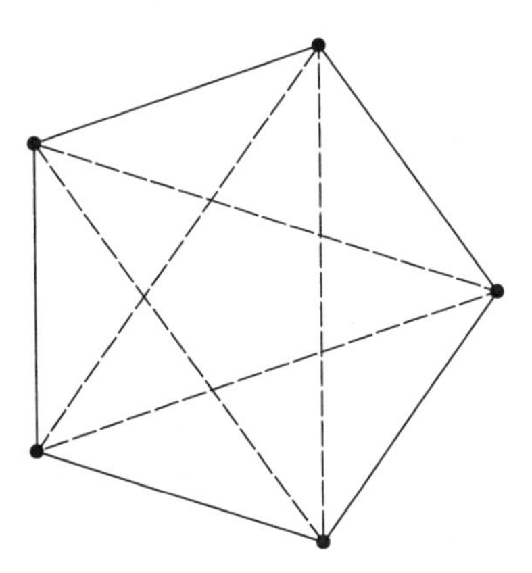

This graph representing the relationships of five people shows that it is possible to obtain a gathering in which there is no group of three strangers or three acquaintances. In other words, there is no triangle in the diagram that is made up entirely of solid or dashed lines.

that four points will be linked by the same color. Only a graph of eighteen or more points does the job. Thus, R(4,4) = 18. In other words, at a dinner party for eighteen there will always be at least four mutual acquaintances or strangers, but that may not necessarily happen when there are only seventeen people present.

It's tempting to consider Ramsey's theorem in drawing up a guest list to ensure an appropriate balance of novelty and familiarity in some random gathering—guests selected, perhaps, by sticking pins in a phone book. The trouble is that calculating the minimum number of partygoers to have—say, seven acquaintances or nine strangers—can be extraordinarily difficult. Though Ramsey's theorem guarantees the existence of such arrangements, determining their actual size is an onerous task.

The chief problem is that the number of cases that must be checked escalates rapidly with each step up in the size of the group. For example, a group of five represents 1,024 different cases, a group of six has 32,768 possibilities, and a group of seven has 2^{21} possibilities.

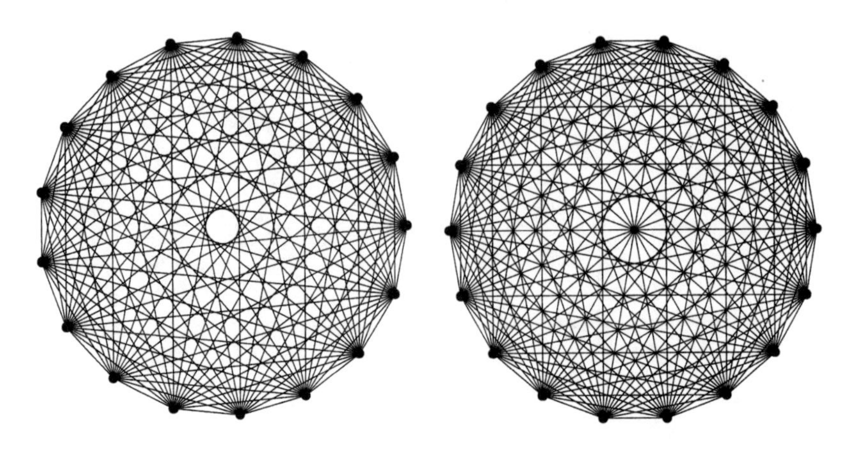

In these graphs, each point represents a person. The lines can be colored so that, say, a red line connects mutual acquaintances and a blue line connects mutual strangers. In a graph representing seventeen people (left), it's possible to color the lines so that no four points are connected by a network of lines that are either completely blue or completely red. In contrast, in a graph representing eighteen people (right), there is always such a network, meaning that a group of four mutual acquaintances or four mutual strangers will always be present in such a gathering.

A prodigious amount of effort involving considerable mathematical ingenuity has gone into producing the rather scant table of Ramsey numbers now known (and listed on the next page). In several cases, mathematicians have been able to find the answer only to within a range of the true value—all for the sake of solving an intriguing puzzle rather than for any particular practical application.

In the last few years, computers have played a significant role in determining Ramsey numbers. In 1990, Brendan McKay of the Australian National University and Zhang Ke Min of Nanjing University in China relied on computers to help them sift through thousands upon thousands of possibilities to determine R(3,8). A little later, McKay began collaborating with Stanislaw P. Radziszowski of the Rochester Institute of Technology in a concerted effort to find R(4,5). They communicated almost entirely by electronic mail, rarely meeting face-to-face.

Checking all possible graphs was out of the question. Instead, McKay and Radziszowski focused on developing efficient procedures for mathematically "gluing" together small graphs, which could be checked more easily, to make graphs large enough for the problem at

Number of Acquaintances	*Number of Strangers*	*Ramsey Number*
3	3	R(3,3) = 6
3	4	R(3,4) = 9
3	5	R(3,5) = 14
3	6	R(3,6) = 18
3	7	R(3,7) = 23
3	8	R(3,8) = 28
3	9	R(3,9) = 36
3	10	R(3,10) = 40–43
4	4	R(4,4) = 18
4	5	R(4,5) = 25
4	6	R(4,6) = 35–41
5	5	R(5,5) = 43–49
5	6	R(5,6) = 58–87
6	6	R(6,6) = 102–165
7	7	R(7,7) = 205–540

A Ramsey number represents the minimum number of people needed to guarantee the presence of a given number of mutual acquaintances and a given number of mutual strangers. In some cases, mathematicians have not yet pinpointed the actual Ramsey number and can give only the range within which the number must fall.

hand. Because their search technique involved using the same computer program many times with different data, the researchers could readily partition their task into small pieces. This meant that they could do the necessary computations on many desktop computers rather than having to rely on a single supercomputer.

Both institutions where the researchers were based had a large number of workstations located in staff offices and student laboratories. Many of these machines were often idle during the night or on weekends. By commandeering these resources, the mathematicians could have as many as 110 computers working simultaneously on the problem. By early 1993, McKay and Radziszowski had their answer: R(4,5) = 25.

At present, the prospect of cracking the next candidate, R(5,5), appears bleak. Even a thousandfold increase in computing power would probably not suffice. Erdös liked to tell an allegorical tale about the difficulties of determining Ramsey numbers: Suppose an evil spirit were to come to Earth and declare, "You have two years to find R(5,5). If you don't, I will exterminate the human race." In such an eventuality, it would be wise to get all the computers in the world together to solve the problem. "We might find it in two years," Erdös predicts.

But if the evil spirit were to ask for R(6,6), it would be best to forgo any attempt at the computation and instead launch a preemptive strike against the spirit to destroy him before he destroys us. On the other hand, "if we could get the right answer just by thinking, we wouldn't have to be afraid of him, because we would be so clever that he couldn't do us any harm," Erdös concludes.

Erdös himself was a master of innovative techniques for solving problems, sometimes even exploiting randomness and probability to establish the existence of structures that he was determined to corral in his many party puzzles.

Dartboard Estimates

Throwing darts at a target may sound like a curiously haphazard way to solve a mathematical problem. Properly applied as a kind of intelligent guessing, however, it can become a highly effective technique for obtaining answers to certain problems in Ramsey theory and in other areas of mathematics.

Suppose that instead of the usual rings and diagonals, a dartboard features a target in the shape of a triangle inside a circle. If one were to throw darts at the board without aiming, the darts would land randomly within the circle and the triangle. By counting up the number of darts that land anywhere on the target and those than land only within the triangle, one can estimate the triangle's size relative to that of the circle. For example, if ten out of a hundred randomly thrown darts fall inside the triangle, the triangle has approximately one-tenth the area of the circle.

Such poorly played games of darts can provide remarkably good estimates of a variety of mathematical quantities, including π (pi), the ratio of the circumference of a circle to its diameter. Just make a target in which a circle of a given radius fits snugly inside a square with sides equal to twice the circle's radius. The number of darts that hit the circle divided by the number that hit the square provides an estimate of the value of π divided by four. The more inaccurate the

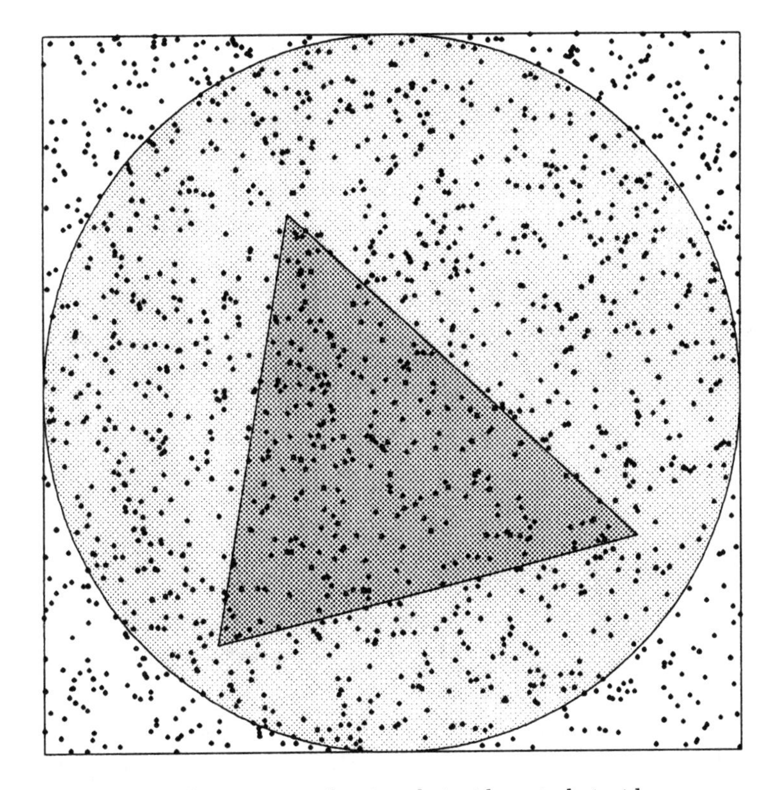

For a dartboard consisting of a triangle inside a circle inside a square, the proportion of randomly thrown darts that land inside the triangle provides an estimate of the triangle's area relative to that of the square. You can also estimate the circle's area relative to that of the square or find an approximate value of π.

aim and the larger the number of throws, the better the estimate becomes.

Random choices can also play a prominent role in mathematical proofs. In 1947, Erdös used a probabilistic method to show, without actually determining the Ramsey number itself, that a given Ramsey number has to be greater than a certain value. The darts of Erdös's thought experiment were flips of a coin. In effect, he started with a room of mutual strangers, brought each of the pairs together, and introduced them only if a coin flip came up heads. The result was a random mixture of strangers and fresh acquaintances, linked in various ways. By an ingenious chain of reasoning, Erdös then demonstrated that the probability of getting a party with a desired mix of strangers and acquaintances is practically certain beyond a particular size of group.

However, although Erdös could prove the existence of the right kinds of patterns among the random pairings of his thought experiment, he could give no hint as to how to find such pairings or draw and color the relevant graphs and subgraphs to determine the Ramsey number itself.

Using mathematical recipes that involve the equivalent of coin flips has proved a useful strategy for solving a wide range of problems in mathematics, from testing whether a number is prime (evenly divisible only by itself and one) to estimating such mathematical quantities as π. Such randomized algorithms are also extremely useful in computer science for sorting data, recognizing patterns, generating graphic images, and searching databases on a computer. In many cases, random choices or paths shorten the task at hand and help to avoid cases of computational gridlock.

The same thing can happen when two people walking in opposite directions along a sidewalk both step to the same side to avoid colliding, then do it again and again. They get trapped in an embarrassing dance until the stalemate is somehow finally broken. A coin toss can solve the problem for both the inadvertent dancers and the computer that is engaged in resolving conflicting instructions or deciding which of two or more equally likely courses to take.

Probabilities also enter into so-called Monte Carlo simulations for modeling the behavior of atoms in a gas and other physical systems (see chapter 9). One can toss coins or throw dice to obtain answers to both mathematical and physical questions, and, with the help of a computer, this can sometimes be done surprisingly quickly and efficiently. Ironically, Erdös himself never worked with a computer. His insights, however, suggested novel probabilistic methods of tackling tough problems well suited to the powerful, ubiquitous computers of today.

Group Thoughts

Mathematical research is generally thought to be a lonely pursuit. One imagines a mathematician squirreled away in a dingy garret, an isolated wilderness cabin, or a sparsely furnished cubicle, thinking deeply, scrawling inscrutable equations across scraps of paper, to emerge from exile at last with proof in hand.

A few mathematicians *do* spend their professional lives in solitary contemplation of a single problem. In general, however, mathematical research is a remarkably social process. Colleagues meet constantly to compare notes, discuss problems, look for hints, and work on proofs together. The abundance of conferences, symposia, workshops, seminars, and other gatherings devoted to mathematical topics attests to a strong desire for interaction.

Paul Erdös, perhaps more than any other mathematician in modern times, epitomized the strength and breadth of mathematical collaboration. Because he had no permanent home and no particular job, Erdös simply traveled from one mathematical center to another, sometimes seeking new collaborators, sometimes continuing a work in progress. His well-being was the collective responsibility of mathematicians throughout the world.

"My brain is open," Erdös typically declared on stepping into a mathematician's office, and the work would begin. For him, doing mathematics was as natural as breathing, and he did it for more than sixty-five years. Driven by the notion that life represents a cosmic struggle to uncover beautiful truths hidden away by a stubborn, contrary God, Erdös applied himself to his pursuit with frenetic zeal. "A mathematician is a device for turning coffee into theorems," he said wryly.

To Erdös, mathematics and its elements were more real than trees and grass, transcending reality and awaiting discovery. At the same time, though he did not like having possessions, Erdös was not an ascetic. He liked to eat well in good restaurants and stay in fine hotels when he got the chance. A compassionate, generous, gentle man, he was well informed on almost any topic of conversation and deeply aware of the tragedies of world politics.

Erdös wrote hundreds of research papers on a wide range of mathematical topics. Especially astonishing was the extent to which he also worked with other mathematicians to produce joint papers. Collaboration on such a scale had never been seen before in mathematics, and it has now entered the folklore of the mathematical community. Of course, there's a characteristically mathematical way to describe this webbiness—a quantity called the Erdös number.

Mathematicians assign Erdös the number 0. Anyone who has coauthored a paper with him has the cherished Erdös number 1. As of March 1997, there were 472 such coauthors. Another 5,016 mathematicians have the Erdös number of 2, because they wrote a paper not with Erdös himself but with someone who wrote a paper with Erdös. People belonging to the first three categories already encompass a significant proportion of all mathematicians in academia worldwide.

The Erdös number of 3 goes to anyone who has collaborated with someone who has collaborated with someone who coauthored a paper with Erdös, and so on. Thus, any person not yet assigned an Erdös number who has written a joint mathematical paper with a person having Erdös number n earns the Erdös number $n + 1$. Anyone left out of this assignment process has the Erdös number infinity.

Keeping track of these mathematical links has become a kind of game, punctuated by published, tongue-in-cheek explorations of the properties of Erdös numbers. Albert Einstein (1879–1955), for instance, had an Erdös number of 2. Andrew Wiles, who in 1994 proved Fermat's Last Theorem, has an Erdös number no greater than 4.

It's possible to draw a collaboration graph in which every point represents a mathematician, and lines join mathematicians who have collaborated with each other on at least one published paper. The resulting tangle is one of the largest, most elaborate graphs available to mathematicians. Some have conjectured that this monstrous graph snares nearly every present-day mathematician and has threads into all areas of mathematics, into computer science and physics, and even into the life sciences and social sciences. Its breadth is astonishing and vividly demonstrates the vital role of collaboration in math and science.

The mathematician Jerrold W. Grossman of Oakland University in Rochester, Michigan, has taken on the task of maintaining a comprehensive, up-to-date listing of mathematicians who have earned an Erdös number of 1 or 2. "It's fun," Grossman says. "But more seriously, it shows that mathematical research is very webby, with almost everybody talking to people who talk to people."

Indeed, fascinating patterns emerge from the study of Erdös collaboration lists. For example, the average number of authors per research article in the mathematical sciences has increased dramatically during Paul Erdös's lifetime. About 50 years ago, more than 90 percent of all papers published were solo works, according to Grossman. Today, barely half of the papers are individual efforts. In the same period, the fraction of two-author papers has risen from less than one-tenth to roughly one-third. In 1940, virtually no papers had three authors. Now, about 10 percent of all papers have three or more authors. Erdös himself may have played a role in this relatively recent explosion of collaboration.

When we see patterns—whether in the arrangement of stars in the sky or in the distribution of guests at a dinner party—we are constantly tempted to think of these patterns as existing for a purpose and being the effect of a cause. Ramsey's theorem suggests otherwise. Patterns can, and indeed must, arise just as well out of pure randomness and pure chance.

In mathematics, in science, and in life, we constantly face the delicate, tricky task of separating design from happenstance. We must make decisions about the meaning of apparent coincidences, whether what we have before us is medical data showing an unusual cluster of cancers in a community, a sequence of events that suggests a conspiracy, or a peculiar arrangement of stars in the sky. We detect eight stars in a straight line and think, That can't be an accident. Is this pattern the result of some cosmic ordering driven by gravity? Is an unknown force involved? Is it a string of deliberately placed beacons for interstellar travel? In the absence of an obvious explanation, the human instinct has been—and still is—to invoke supernatural forces, extraterrestrial visitors, or other fantastic beings and events to make sense of what we observe.

Humans are model builders, and the human mind is very good at identifying patterns and constructing theories. We are programmed to search for patterns and to invent explanations, and we find it difficult to imagine that patterns emerge from randomness.

As builders, humans also create edifices on a vast range of scales, from the giant pyramids of ancient times to the intricate microscopic components of a computer's microprocessor. We can manipulate individual atoms to spell out tiny words on a silicon surface, and we can orchestrate the construction of giant skyscrapers and vast malls. To do so, we design, make plans, create blueprints, organize labor, and marshal resources to create the order that we desire. In nature, however, that order and structure arises from randomness, and we face the puzzle of how the components of life assemble themselves without blueprints as guides.

3

Shell Game

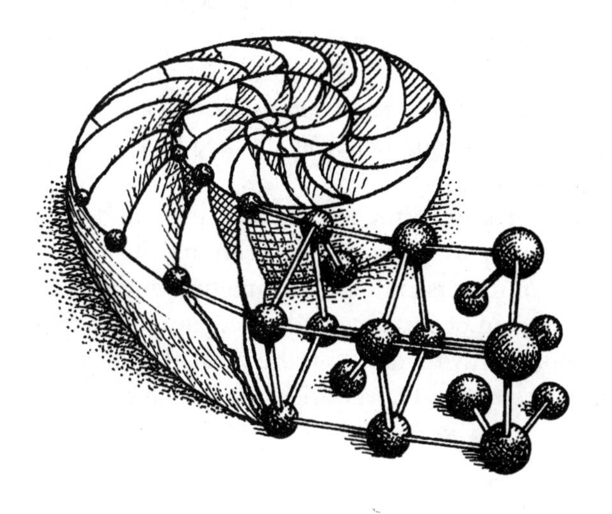

When we mean to build,
We first survey the plot, then draw the model,
And when we see the figure of the house,
Then must we rate the cost of the erection.

—*William Shakespeare (1564–1616),* King Henry IV, Part II.

Behind a tall, sturdy hoarding encroaching upon a cracked sidewalk, a maze of steel girders and concrete slabs begins to assume the shape of an office building. Workers pound nails, weld joints, pour concrete, and shift heavy blocks into place. To one side, an engineer sporting a hard-hat consults a wide sheet unfurled from a roll of blueprints. Another supervises the erection of a wall.

When nature builds, there's no architect or engineer present to plan and supervise the project, whether it's a living cell or a mature tree. There's no one to organize the effort or check the results. Nature's structures assemble themselves from building blocks available in the environment, captured from a roiling sea of material and set into place as required. Order emerges out of randomness.

Atoms combine into molecules, molecules assemble themselves into the components of a living cell, and cells join together into complex organisms. At each stage, the structures grow more intricate and varied. At the heart of this frenetic construction are complicated molecules called proteins and nucleic acids, the chief concerns of molecular biology.

The mathematics of shapes and structures and of combinations and graphs represents a means of discerning the rules of biological structure and of understanding how function emerges from what appears to be random occurrence and arrangement. Such a mathematical framework is proving useful in the escalating arms race that medical science faces against adversaries as tough and versatile as viruses.

Cough Machine

The subway car is crowded with commuters heading downtown early on the morning of a wintry day. The train's chorus of rumbles, rattles, and screeches drowns out the shuffling of feet adjusting to the car's sway and jerk, the rustle of newspapers, and the mutter of conversation.

A passenger pulls out a handkerchief to wipe a runny nose. Bending over a book, another sniffles. A third intercepts a raspy, throat-clearing cough, using a thick sleeve to muffle the outburst. Someone else abruptly sneezes, explosively venting a spray of fine droplets into the air.

The sudden, noisy expulsion of air from the lungs is a defensive reflex designed to keep the air passage free of irritating material. Specialized sensors detect the presence of a specific threat and send the data to the brain. This signal activates a cough center, which in turn triggers an elaborate set of coordinated movements involving the diaphragm, the chest muscles, and the voice box. These actions propel mucus—a protein slime laced with trapped foreign matter—up the windpipe to the back of the throat, where it can be spit out or swallowed and sent to the stomach to be destroyed by acid.

Bacteria and viruses are among the irritants that can trigger such a response. A viral respiratory infection can provoke the discharge of so much mucus that a sufferer can easily go through an entire box of tissues in one day. Viruses infect a wide range of plants, animals, and bacteria. In humans, rhinoviruses are responsible for the common cold. The poliovirus causes infantile paralysis. Other viruses inflict such diseases as rabies, hepatitis, German measles, yellow fever, smallpox, mumps, and acquired immune deficiency syndrome (AIDS).

Like a microscopic, heavily armored spacecraft, a virus in the human body cruises intercellular space in search of a cell to invade and colonize. A tough, rigid shell protects its cargo of nucleic acids—molecular strands of genetic material. Making contact with a cell wall and clinging to its target, the virus shell opens up to inject its contents into the hapless cell.

The success of a virus in infecting its host depends largely on how well the microbe's outer coat functions. Made up of protein building blocks, this tightly closed shell usually has a highly regular structure. As the virus reproduces itself inside the cell it has infected, new shells must assemble themselves out of a broth of proteins.

In recent years, a small group of mathematicians has pioneered a novel perspective on virus self-assembly—how structural order emerges out of randomness in the microcellular realm. This research suggests that sets of simple rules, which define the way proteins stick together, automatically lead to the kinds of virus structures that biologists observe under their electron microscopes.

Typically less than 0.3 micrometers in size, viruses have highly regular structures; often they look like mineral crystals with flat faces, distinct angles, and definite edges. In fact, a large proportion of all known viruses resemble the twenty-faced, three-dimensional geometric figure known as an icosahedron.

The icosahedron is one of only five three-dimensional forms whose faces are made up of regular polygons. The other four are the tetrahedron, which consists of four equilateral triangles; the cube (or

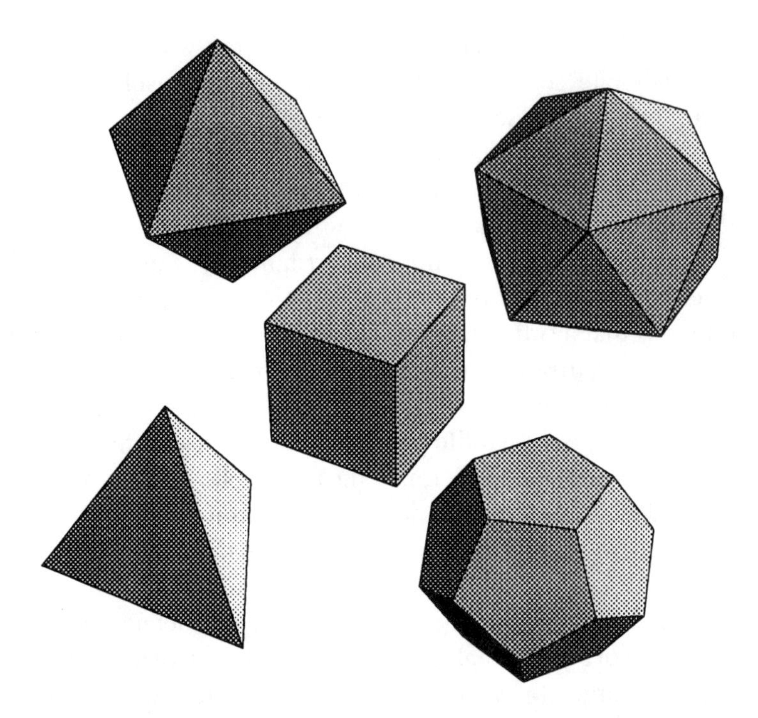

The five Platonic solids are the only ones that can be constructed from a single type of regular polygon: four, eight, or twenty equilateral triangles in the case of the tetrahedron (lower left), the octahedron (upper left), and the icosahedron (upper right); six squares in the case of the cube (middle); and twelve pentagons in the case of the dodecahedron (lower right).

hexahedron), which has six squares as faces; the octahedron, which is made up of eight equilateral triangles; and the dodecahedron, which has twelve pentagons.

The Pythagoreans of ancient times studied and revered the five geometric forms, ascribing to them mystical properties. The Greek philosopher Plato (427–347 B.C.) featured them prominently in his writings about the constituents of the universe.

In Plato's view, the four elements that made up the world—fire, air, water, and earth—were all aggregates of tiny solids, each one having the shape of one of the regular polyhedra. As the lightest and sharpest of the elements, fire was a tetrahedron. As the most stable, earth consisted of cubes. As the most mobile, water was an icosahedron, the regular solid most likely to roll easily. Air had to be an octahedron, while the dodecahedron represented the entire universe.

The last proposition of Euclid's *Elements*, a compendium of geometric knowledge, is an elegant proof that these five forms—often called the Platonic solids—are the only ones that can be constructed from a single type of regular polygon.

What is it about an icosahedron that might make it the preferred shape of a virus? As shown by the shape of a soap bubble, a sphere represents the most efficient way of enclosing a given volume of space using the least amount of material. For a closed shell constructed from a relatively small number of identical units, the icosahedron is the regular solid that most closely approximates a sphere.

The structure of a typical virus resembles that of an egg. It has a protective shell that surrounds a core of genetic material consisting of loops or strands of deoxyribonucleic acid (DNA) or sometimes ribonucleic acid (RNA). A DNA molecule looks like a ladder twisted into a spiral, with each rung consisting of a pair of much simpler molecules called bases. The sequence of bases along a strand encodes instructions for constructing an array of proteins.

A virus shell is made up of proteins, which are large molecules composed of amino acids. A single protein molecule, however, isn't big enough to encompass all the genetic material that a virus typically carries. Many protein units must come together to form the complete shell. The resulting virus shell, or coat, is called a capsid. Sometimes this shell is further encased in a flabby envelope of fatty material. The entire structure adds up to a highly compact vehicle for delivering a potentially deadly message to a living cell.

Geodesic Numbers

One of the most striking attractions of Expo '67, the world's fair held in Montreal, was the enormous geodesic dome that housed the United States pavilion. Designed by architect, engineer, and visionary R. Buckminster Fuller (1895–1983), the dome consisted of self-bracing triangles of metal struts. The result was a wonderfully strong, light, airy structure that seemed to defy gravity—a bubble barely tethered to the ground.

Built in mere hours rather than the weeks or months required for conventional structures, geodesic domes have served as shelters atop mountains and as enclosures for Arctic radar installations. They have survived earthquakes and hurricanes. Yet their skins are proportionally

Buckminster Fuller was famous for his geodesic dome designs. (From J. Baldwin, *BuckyWorks*, © 1996 John Wiley & Sons.)

thinner than a chicken's egg shell is to the egg, and they require no internal supporting skeleton.

Fuller saw the same geometry that he employed in his geodesic domes in the structure of viruses. "Although the poliovirus is quite different from the common cold virus, and both are different from other viruses, all of them employ [geodesic patterns] in producing those most powerful of structural enclosures of all the biological regeneration of life," he wrote in 1975 in *Synergetics*. "It is the structural power of these geodesic-sphere shells that makes so lethal those viruses unfriendly to man. They are almost indestructible."

Virus shells come in a variety of sizes, depending on the type of virus. Each virus has a unique set of coat proteins, whose shape and composition dictate how the identical units fit together. In each case, the completed shell must be big enough to hold all the genetic material necessary for that particular virus.

The icosahedral shell of the satellite tobacco necrosis virus is made up of sixty protein units. There is a protein at each corner of each of the shell's twenty triangular faces. One can also imagine the structure as being made up of five proteins gathered at each of the twelve corners, or vertices, of an icosahedron, as is shown in the figure below. In effect, its surface can be thought of as consisting of twelve protein pentagons.

Larger shells have additional protein units at the corners of their triangular faces. For example, the poliovirus shell consists of 180 coat proteins, with three proteins in each corner, for a total of nine on each face. This structure can also be pictured as consisting of twelve groups of five proteins each at the twelve vertices of an icosahedron and twenty groups of six proteins each at the center of each of the faces. The overall structure resembles the patchwork pattern of a soccer ball sewn together from twelve pentagons and twenty hexagons of leather.

Indeed, electron-microscope images of virus shells don't usually show clearly defined triangular faces. Instead, one sees lumpy protein globs arranged as arrays of hexagons and pentagons. Such striking geometric features naturally led biologists to assume that a virus shell's final

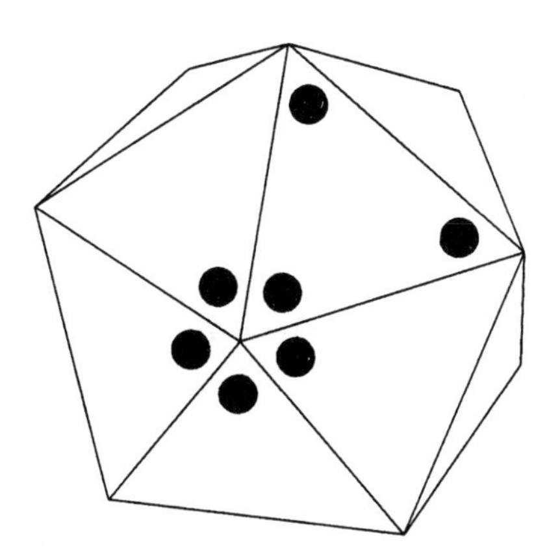

An icosahedron has twelve vertices, twenty triangular faces, and thirty edges. A virus shell constructed from sixty proteins has three proteins on each triangular face. You can also view the sixty proteins in groups of five surrounding each of the twelve vertices of the icosahedron.

shape directly reflects the geometry of the protein building blocks and the way they link together to create a shell.

In 1962, D. L. D. Caspar and Aaron Klug postulated that all virus shells that resemble icosahedra are pieced together from protein blocks in the shape of pentagons and hexagons. In other words, five or six coat proteins initially link to form a pentagon or a hexagon, and these pentagonal and hexagonal units fit together to create an icosahedral structure.

Inspired by the geometry of the geodesic domes created by Buckminster Fuller, Caspar and Klug built their virus structures out of a framework of triangles. Suppose a wide, flat area is covered by equilateral triangles fitted together neatly with no spaces between them. Such an arrangement is known as a tiling of the plane. A unit containing six of these triangles constitutes a hexagon, so this triangular grid, as shown in the figure below, is also a honeycomb tiling of hexagons.

A sheet of hexagons is flat, so there's no way to create a closed shell out of just hexagons. Every once in a while, however, one of these hexagons can be replaced by a pentagon, which has one fewer side than a hexagon. Placing pentagons among the hexagons causes the flat sheet to curl into a three-dimensional object.

It turns out that to get an approximately round object, a particular relationship must hold between the number of pentagons and hexagons present. The formula underlying that relationship was discovered by the Swiss mathematician Leonhard Euler (1707–1783), who with 886 books and papers to his credit, was the most prolific writer of mathematics in history. His eighteenth-century contemporaries called him "analysis incarnate" for his remarkable mathematical prowess.

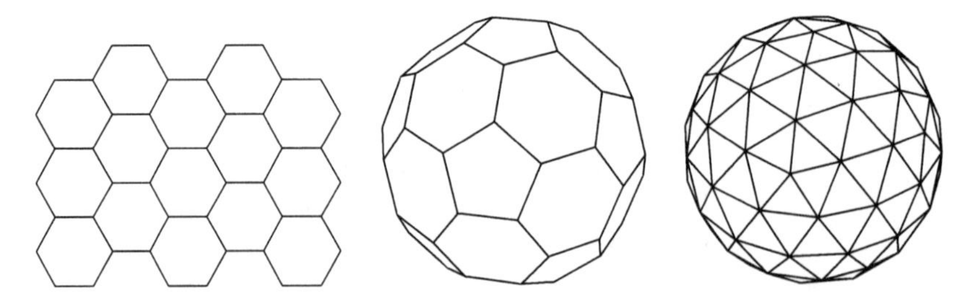

A honeycomb tiling of hexagons produces a flat surface (left). An arrangement of precisely twelve pentagons and twenty hexagons creates a closed structure resembling a soccer ball (middle). These polygons, in turn, can be rendered as sets of triangles to create a geodesic structure (right).

In 1750, Euler wrote a letter to the mathematician Christian Goldbach (1690–1764), in which he mentioned a remarkable formula he had discovered linking the number of vertices, edges, and faces of a polyhedron: If F is the number of faces of a closed, three-dimensional figure, V the number of vertices (or corners), and E the number of edges, then $F + V - E = 2$. Applied to a cube, which has six faces, eight vertices, and twelve edges, the formula gives $6 + 8 - 12 = 2$. For a tetrahedron, which has four faces, four vertices, and six edges, $4 + 4 - 6 = 2$. For an icosahedron, with twenty faces, twelve vertices, and thirty edges, $20 + 12 - 30 = 2$.

In the case of icosahedral polyhedra, the formula works only if precisely twelve pentagons are present along with the requisite number of hexagons for each different size of shell.

These assumptions limit the number of possible, roughly spherical shell structures, each one containing twelve pentagonal units and a certain number of hexagonal units. To catalog the possible geometries, Caspar and Klug defined a number, T, which corresponds to the number of coat proteins at each corner of a triangular face of the shell. Thus, $T = 1$ for the shell of the satellite tobacco necrosis virus, and $T = 3$ for the poliovirus shell.

In this virus shell model, the only T numbers allowed are 1, 3, 4, 7, 9, 12, 13, 16, 19, 21, 25, and so on. For each type of icosahedral shell, the number of faces is $20T$, the number of edges $30T$, and the number of vertices $10T + 2$, as you can see in the following table.

T	*Number of Vertices ($10T + 2$)*	*Number of Protein Units ($60T$)*
1	12	60
3	32	180
4	42	240
7	72	420
9	92	540
12	122	720

The T value represents the number of coat proteins at each corner of a virus shell's triangular face. From this value, you can derive the number of vertices in the corresponding shell's icosahedral shape as well as the total number of proteins present.

Fuller noted the great stability of structures based on the "magic" $10T + 2$ numbers—12, 32, 42, 72, 92, 122, and so on—not only in his geodesic dome designs but also in viruses. "I made that discovery in the late 1930s and published it in 1944," he wrote in *Synergetics*. "The molecular biologists have confirmed and developed my formula by virtue of which we can predict the number of nodes in the external protein shells of all the viruses, within which shells are housed the DNA-RNA-programmed design controls of all the biological species and of all the individuals within those species."

Based on the belief that icosahedral virus shells are built out of hexagonal and pentagonal protein units, the Caspar and Klug formula provides a convenient way of classifying shells according to their T number. This model, however, doesn't really address the question of how individual proteins join together to create the pentagons and hexagons that form a virus shell. How does a coat protein, wandering about in a thick soup, know where to go without a blueprint to guide it?

Virus Factory

A bacteriophage is a virus that infects only bacteria. The *T*4 bacteriophage looks like an icosahedron sitting atop a pedestal on the back of a spindly-legged spider that's missing two of its limbs. When this virus lands feet first on a host cell, its protein-fiber legs anchor themselves to the bacterial cell wall. The collapsible hollow pedestal acts like a syringe to drive the DNA genetic material out of the virus into the bacterium. The empty protein shell stays behind on the cell's surface and eventually disintegrates. The P22 bacteriophage, a cousin of the *T*4, has a simpler structure and a shorter pedestal, but it is just as effective in delivering its lethal cargo.

The injected DNA then commandeers the metabolic machinery of its host cell, directing the creation of more viral DNA. This activity leads to the production of viral protein, including the material needed to construct new shells. Finally, the bacterial wall bursts, releasing hundreds of factory-fresh bacteriophages.

The P22 bacteriophage is one of the most intensively studied of all viruses, and the biologist Jonathan King of the Massachusetts Institute of Technology has recently devoted much of his time to determining precisely how its coat proteins assemble into an icosahedral structure. Experimental data obtained by King and Peter E. Prevelige, Jr., presently a microbiologist at the University of Alabama, indicate that assembly occurs very rapidly, generally in a matter of seconds, making it extremely difficult to monitor intermediate steps. These studies also reveal the presence of individual proteins and completed shells, but not the pentagonal and hexagonal units that have been postulated as crucial intermediate components in the building process.

In some instances, virus shells require a protein scaffolding to aid assembly. This disposable framework serves the same stabilizing function as the scaffolding needed to support a cathedral's dome or a bridge's arch during construction, and it disappears once the structure has been completed.

King and Prevelige's experimental results pose a challenge to the conventional model that shells form out of protein hexagons and pentagons. The symmetry and regularity of the end product are apparently not a good guide to elucidating the actual building process.

Several years ago, King became interested in using computers to visualize in three dimensions the ways in which protein shells might grow. He suggested the problem to Bonnie Berger, then a postdoctoral student with a background in both computer science and mathematics. He thought that she might be able to come up with a computer program that would make it possible to depict protein polymerization—the joining together of proteins to form larger units—on the computer screen.

Berger, now a professor of applied mathematics at MIT and a member of its Laboratory for Computer Science, had developed a strong interest in the application of mathematics to problems in biology, especially the question of how long, stringy protein molecules fold themselves into compact globs. At that time, she knew very little about viruses, but King's problem sounded intriguing.

Berger ended up describing the problem to Peter Shor, a mathematician at Bell Labs, with whom she had worked previously. They began exploring the kinds of rules that might specify how one protein links up with another to create a certain structure. Their task resembled that of solving a three-dimensional jigsaw puzzle, except that they also had to fabricate the pieces, then check whether these pieces linked properly to produce the required structure.

Berger and Shor initially focused on the icosahedral shell of the P22 bacteriophage, which has a *T* number of 7 and consists of 420 protein units. They knew that each coat protein had three links to neighboring proteins. To distill the problem to its essence, they drew a graph—a set of nodes (representing proteins) joined by lines (indicating binding interactions)—to illustrate how the proteins fit together in a P22 virus shell, as is shown in the figure below.

The mathematicians didn't make much progress in formulating linking rules, however, until they realized that such schemes could be made to work if the protein units, though identical in chemical composition, actually had several different shapes, or conformations. At the time, Berger didn't know whether coat proteins actually have this property, but assuming that they did made it possible to show why some proteins link together into pentagons and others become part of hexagons.

For the P22 bacteriophage, Berger and Shor worked out a scheme based on the use of seven different building blocks, each corresponding to a glob of protein in one of seven possible conformations. It turned out that only one protein conformation was needed for the groups consisting of five proteins, and six conformations were needed to form the hexagonal arrangements.

The mathematicians numbered the different building blocks from 1 to 7 and deduced linking rules for each type. For each conformation, the combinatorial part of the rules specified which three blocks it could fit with. For example, the type 1 conformation could link with two other type 1 blocks and with a type 2 block. A type 2 block could stick to a type 1, 3, and 7 block, and so on.

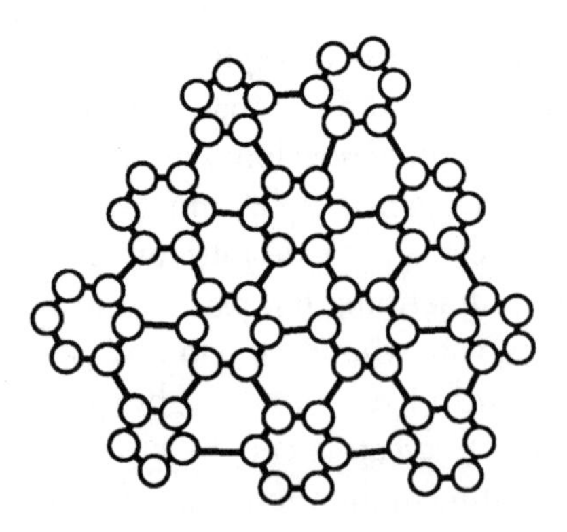

The arrangement of proteins in a virus shell can be represented as a graph in which the proteins are circles and the bonds between them are lines. (Courtesy of Bonnie Berger, MIT)

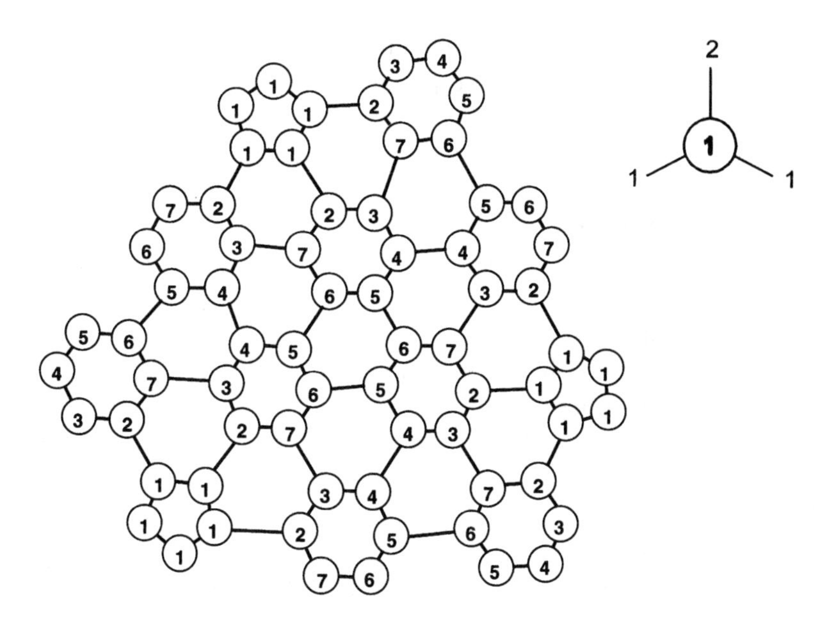

Assembling a P22 virus shell is a bit like solving a matching puzzle. There are seven protein shapes and seven rules specifying which protein shapes can link together. For example, protein 1 forms three bonds—two with proteins in conformation 1 and one with a protein in conformation 2. The trick is to number all the circles in the graph according to the rules so that everything fits together. (Courtesy of Bonnie Berger, MIT)

Berger and Shor also found that they could formulate an alternative set of rules, involving just four different protein conformations, and still obtain an icosahedral structure. When Berger checked with King, she was delighted to learn that he and his coworkers had recently obtained experimental evidence strongly suggesting that a protein's shape could depend on its location within a shell. King, in turn, was astonished that Berger had come up with precisely the same number of conformations that his experimental studies had uncovered. His electron micrographs showed that the protein molecule that makes up the P22 shell has seven different shapes, of which four are particularly distinctive. In other words, the same chemical units can exist in different, stable conformations when they are in different positions in a structure. They behave like floppy Legos, which can reshape themselves in specific ways to fit into particular places in a model structure.

Berger and Shor had discovered a plausible model for shell assembly, one that required no blueprints or special mechanisms to direct each protein to its proper place in the virus coat. It's like covering a floor with hexagonal tiles. The person who's laying the tiles doesn't need to know the overall pattern or precisely where every tile goes at the outset. The floor pattern emerges by simply matching the sides of each new tile with others that have already been laid down. Local rules generate the structure.

In a cell, a pair of protein molecules floating about randomly would collide and lock together. Once those two were linked, a third could come along, make contact, and twist itself into just the right shape to fit in and join the assemblage. Step by step, the structure would grow into its icosahedral form.

To test the model, Berger and several of her coworkers developed a computer simulation in which randomly moving particles, representing protein molecules, collide and stick together according to the rules. The trial simulations for the $T = 7$ case demonstrated convincingly that repeated application of the rules indeed created closed shells of the proper geometry (see color plate 1).

The researchers also discovered that starting shell formation with the wrong initial protein cluster produced a malformed shell that couldn't close on itself. Instead, the protein shell would keep growing, creating an ever-enlarging spiral structure. This occurred when proteins accidentally joined into a hexagonal arrangement rather than a pentagonal structure at the beginning of shell formation. Thus, with one false step, a polymerizing shell could start growing into a spiral, producing a structure like that of some seashells.

Biologists have, in fact, observed and identified a number of types of malformed virus shells, including spiral and tubular structures. Berger's local rule theory suggests that these arise from mistakes in the assembly process, when otherwise normal proteins fail to interact and link properly.

Berger and her colleagues have recently developed new computer simulations that model the rates at which different types of shells grow. Instead of proteins locking firmly into place according to the rules, they can come and go, sometimes bonding and sometimes not. The researchers can also vary the physical properties of the particles, changing their size, shape, mass, bond strength, and bond length to determine what role such factors may play in shell-protein interactions and to illuminate how the given rules function under different circumstances. The new simulations also allow a protein to switch randomly among different conformations as it wanders about in the more realistic environment of a broth of free-floating particles.

Using computer simulations, the research team can try out different construction scenarios to see more clearly what factors significantly influence and direct virus shell growth. Such tests of the model lead to insights into the type of particles or conditions that might interfere with shell growth, providing data that can be checked against experimental results, to refine further the simulations and theoretical models.

The rules of Berger's model represent one way of expressing the fact that proteins tend to fold themselves in different ways in different settings. Each protein in the assembling network has enough information to form a shell that is precisely the right size for its intended genetic cargo. The new model of shell formation also suggests novel strategies for treating viral infections—methods that hinge on keeping shells from forming properly.

Viral Concoctions

Some viruses are so resilient that they can survive a temperature of −100°C. Some are strong enough to withstand a crushing force a hundred thousand times greater than gravity. Certain viruses can reawaken after hundreds of years of being in suspended animation. Others, like the AIDS virus, are so fragile that they quickly disintegrate when exposed to air or water.

Different viruses wage their campaigns of survival in different ways, seeking out cells that match their needs. A hepatitis virus, for instance, targets liver cells, and a cold virus aims for the cells of your nose. The rabies virus attacks nerve cells. Yet, powerless to move on their own because they lack fins, tails, or any other means of propulsion, these agents depend entirely on chance to find their hapless hosts. They simply float until they happen to reach a congenial resting place.

The human body reacts to such intruders by mobilizing special "killer" cells in its immune system to destroy infected cells and the viruses within them. It also produces antibodies—proteins in the blood or on the surface of certain cells that recognize approaching viruses, stick to them, and render them harmless. Vaccines work by stimulating the production of antibodies against particular viral invaders.

These defensive strategies are not always sufficient, however, and biomedical researchers have tried a variety of approaches to short-circuit the activities of potentially lethal viruses. Typically, such intervention has focused on disrupting the replication of the nucleic acid of a virus inside a cell, preventing a mature virus from binding to a cell, or interfering with some other stage in the infection process. But little attention has been paid to the shell-formation stage, mainly because of the absence of both theoretical and experimental work demonstrating the feasibility of such an approach.

The shell-formation model developed by Berger and her colleagues suggests a novel and potentially lifesaving strategy for disrupting the infection process. Instead of looking for ways to prevent a fully formed shell from binding to a cell, biomedical researchers can begin to think about interfering with the growth of a virus shell to render it ineffectual. In other words, one can try to poison the shell complex of a growing virus by causing a deformity or keeping it from closing.

One group of researchers has already demonstrated that a dose of small dye molecules can block virus shell formation in vitro by binding with shell proteins. That's quite surprising, because these proteins have large surface areas and the dye molecules are comparatively small. The dye molecules apparently zero in on the same molecular sites that, according to the mathematical model, dictate the binding of proteins to one another. Thus, it's possible to stop protein reactions with small, weakly bound molecules.

Such a strategy can be refined further. Guided by a mathematical model of how virus shells may grow, researchers can determine precisely where and when to intervene at the molecular level to disrupt the viral reproductive cycle. The preferred course is not to go after individual proteins but to rechannel the growth into misshapen viral forms that can no longer function properly.

In 1996, Berger and Prevelige proposed the development of a new class of antiviral agents that interfere with the way proteins assemble into a viral shell. They suggested that such a method could reduce the time it takes to develop drugs to combat newly emerging viruses. "The pieces of a capsid [shell] fit like a jigsaw puzzle," Prevelige says. "If an antiviral agent binds to a knob of a puzzle piece, it blocks assembly of the puzzle." One factor that makes this an attractive strategy is that it would work for any virus, although the specific antiviral agent would differ from one virus to another.

Berger and Shor have also extended their mathematical theory of virus shell formation beyond the $T = 7$ case. Their mathematics of linkage generates all the structures in the Caspar and Klug classification of virus shells according to T number.

The mathematicians discovered that it was always possible to reproduce a given structure by formulating as many rules as the T number for that structure. Thus, for $T = 4$, they started with four different protein conformations and worked out four rules to specify their interactions. The $T = 13$ case would require thirteen such conformations and rules. Moreover, Berger and Shor proved mathematically that a given set of rules for an icosahedral structure guarantees the final form.

In a number of situations, alternate sets of rules requiring fewer kinds of protein building blocks can also produce the same result as a full complement of rules. In the $T = 7$ shell, for example, it's possible to use four conformations instead of seven. The existence of this alternate set of rules suggests a close relationship between $T = 4$ and $T = 7$ structures. In fact, biologists have already found evidence of such a connection. For example, although the P22 virus normally produces $T = 7$ shells, it can sometimes create $T = 4$ shells instead.

Berger's theory also predicts the existence of shell geometries that the old biological model of virus shell construction appeared to rule out. For instance, there's no compelling reason that shells corresponding to the T numbers 2, 5, 6, and so on shouldn't exist. Biologists have begun to identify examples of these so-called anomalous forms, which don't fit the usual classification scheme based on the Caspar-Klug theory. The polyoma virus, which causes cervical cancer, belongs to this category.

Despite the predictive power of Berger's rules, they have limitations in their direct application to living systems. These interaction rules say little about what physical characteristics proteins must have in order to behave appropriately. They provide no information about how quickly shells form and in what order the building blocks assemble. Indeed, biologists have only scant evidence that a virus shell assembles itself according to any set of such rules—but, until now, they hadn't thought to look for signs of these particular mechanisms.

So, a new mathematics of viruses that depends on connections and combinations rather than on purely geometric considerations is now in place. It's up to biologists to see how well it applies. "There are many problems we face as biologists where we need the tools of the mathematician to come to our aid," King remarks. "The intersection of biology and mathematics can be extremely productive when both sides tune in to the actual problem, the actual need."

Folding the Sheets

Proteins truly represent the fabric of life. The human body manufactures about a hundred thousand distinct proteins, including those that make up hair, skin, and cell walls; messengers for carrying nutrients, such as the hemoglobin of red blood cells; and enzymes that catalyze chemical reactions, including the digestion of food.

All these proteins, along with those that make up virus shells, consist of head-to-tail joinings of smaller components called amino acids. There are just twenty types of amino acids, which act like a twenty-letter alphabet when strung together to spell out a particular protein. However, it is the way that an individual protein strand bends and twists itself into a ragged ball—a particular three-dimensional shape—that defines what it does.

One of the greatest challenges now facing mathematicians who are interested in biology is a type of molecular origami: predicting from its molecular arrangement the three-dimensional configuration into which a protein folds itself, whether in hemoglobin or a virus-shell building block. The question of how loosely coiled proteins crumple into specifically shaped balls able to perform crucial tasks in a living cell is a major unsolved problem that brings together many aspects of physics, chemistry, biology, and computational mathematics.

Understanding how proteins fold and being able to predict a protein's final shape from its sequence of amino acids would give scientists the power to design new therapeutic agents and to help unmask the role of newly discovered genes, which regulate the construction of proteins. Indeed, getting proteins to fold properly is an important part of designing new products, including sensitive, reliable diagnostic tests for various ailments and drugs that are free of side effects.

Modern biotechnology has also made it simple to determine the amino acid sequences of huge numbers of proteins. At the same time, finding the structures of proteins remains a formidable challenge. Hence, researchers know many more protein sequences than protein structures. Yet it's the structures that decide what proteins actually do.

In principle, the laws of physics determine how a protein folds. Attractions between different portions of a strand pull it into a compact arrangement. In trying to model such a process, however, even a short string of amino acids can fold into such a large number of different configurations that it becomes difficult to sift through the possibilities to find a stable arrangement.

Fortunately, the folding problem can be simplified. The twenty amino acids fall into two distinct groups according to whether they attract or repel water molecules. Most cells are 70 to 90 percent water. The attractive and repulsive forces drive protein folding, creating globular structures in which water-repelling amino acids end up on the inside and water-attracting amino acids on the outside.

At a very simple level, one can imagine a protein as a jointed structure—a stiff strand consisting of two types of beads—that can snap into a small number of different positions at each joint. In developing a workable computer model, one can start with a strand only twenty-seven units long, which would fit snugly when folded into a cube three units wide. It is assumed that each fold creates a right angle, and the entire folded strand stays within the confines of the cube.

Even in this simplified case, with just two types of beads representing amino acids, there are 2^{27} possible sequences and 103,346 different folded structures to check for each sequence. In more realistic cases, the possible number of sequences can be enormously larger: 20^{400} possible

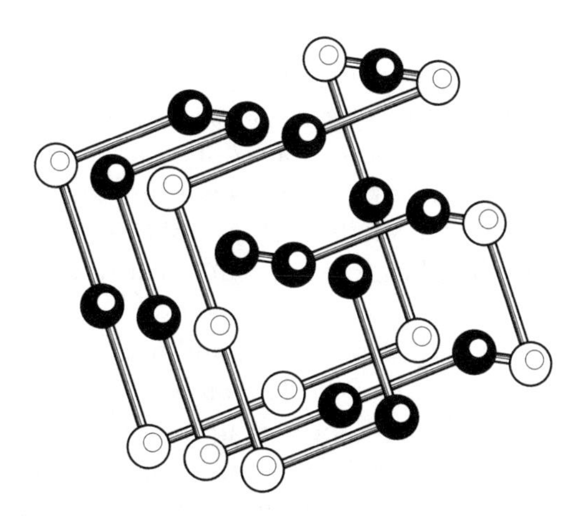

Imagine a protein as a stiff, jointed strand of beads that can snap into various configurations. Black beads represent water-repelling amino acids, and gray beads correspond to water-attracting amino acids. A strand consisting of twenty-seven beads can fold into a compact 3-by-3-by-3 cube, with mainly water-attracting amino acids on the outside. (Courtesy of Chao Tang, NEC Research Institute)

proteins, each four hundred units long, constructed from twenty amino acids. Only a tiny fraction of these possible sequences actually occurs in nature, however, which raises the intriguing question of why these particular ones were favored by evolution. Do they exist because they fold easily or because they represent simple designs stable against mutations in the sequence? That issue remains unresolved.

Another approach to protein folding is to look at it in stages, starting with a floppy chain. As time passes, the wiggling chain becomes increasingly compact. It first settles into large-scale structures, in which some sections twist into helices and others fold back and forth to form sheets. Additional coalescence and folding adds finer structural details, ultimately leading to a roughly spherical shape.

However, there is little agreement, as yet, on the nature and number of folding pathways that lead to a given final structure. Nonetheless, some trends are now readily apparent. For example, when the same sequence of amino acids repeats itself throughout a protein chain—giving the protein a high degree of regularity—the result is often an elongated, fibrous structure, which is commonly found in hair, skin, cartilage, and silk.

It may turn out that nature prefers certain types of folds and arrangements. Because these special structures would probably constitute only a tiny fraction of all possible configurations, finding such patterns would make the task of structure prediction simpler by greatly restricting the number of possibilities that must be examined.

Indeed, researchers have already found that certain structural motifs occur again and again among the proteins now cataloged. Some biologists estimate that only about a thousand types of folds exist among the structures of natural proteins. "Research into protein folding . . . is proceeding enormously faster today than in the past," the biochemist Frederic M. Richards of Yale University noted several years ago. "Those of us involved in the effort still cannot 'play the music,' but we are rapidly learning certain of the notes. That progress alone is heartening, as is knowing that a solution to the folding problem will resolve a question of deep scientific interest and, at the same time, have immediate application in biotechnology."

For mathematicians, it's all part of calculating the secrets of life.

Life, in all its messiness and detail, represents a triumph of order over randomness. It constitutes pockets of structure and purpose distilled from brews of endlessly jostling components. Mathematics serves to illuminate the dynamic crystallization that brings these pieces together and gives structure to proteins—and to life.

A deeper puzzle, however, is that of the evolution of the varied structures of life, from proteins and nucleic acids to cells and complex organisms. How significant a role did chance play in the origin of life on Earth, in the formation of the first amino acids, nucleic acids, and proteins, and in the development of complexity in the myriad life-forms that have lived on Earth at one time or another?

There's an additional complication. Biological structures are not static; protein molecules, for example, are never at rest. They interact with their neighbors and their environment. Complex organisms also move, breathe, and eat. They display coordinated activity. Like the world of structures, the realm of movement has its own patterns that emerge out of randomness.

In all its complexity, life requires both stability and change.

4

Call of the Firefly

Now sleeps the crimson petal, now the white,
Nor waves the cypress in the palace walk;
Nor winks the gold fin in the porphyry font:
The fire-fly wakens: waken thou with me.

—*Alfred Tennyson (1809–1892)*, The Princess: A Medley

Like a flitting spark above a campfire, a firefly glows for a moment before fading to black. In the deep shadows, another living spark briefly announces its presence, then another and another. Each firefly signals at regular intervals; together, they create a magical nightscape of intermittent flashes.

Delightful light shows presented by flying beetles of the family Lampyridae are a familiar summer sight over lawns and meadows across North America and Europe. The performances of fireflies that inhabit mangrove trees along the swampy banks of tidal rivers in Southeast Asia are even more spectacular. At dusk, thousands upon thousands of males gather on the leaves of the trees. Their flickerings are at first completely uncoordinated, but as time passes the flashes drift into synchrony. Within an hour, the entire swarm is flashing on and off in unison. Their combined light is so bright that fishermen out at sea can use the flashing throngs as navigational beacons to guide them back to their home rivers.

An 1854 account of the phenomenon noted: "It is a magnificent spectacle to see spring out at one time from all the branches [of a great tree on the bank of the river], thousands of great electric sparks. . . . It is difficult to explain how this emission of light is simultaneous for several thousands of individuals." The synchronized flashing of these Asian firefly species presents an intriguing puzzle. How do fireflies, cued only by their immediate neighbors' lights, coordinate their signals?

Such synchronization is also evident in the chorusing of crickets, the coordinated beating of pacemaker cells in the heart, and the brain-based electric signals that control such rhythmic behaviors as breathing, running, and chewing. All of these situations involve oscillators that must interact in some way to beat with a single rhythm.

To understand diverse systems in which individuals—whether they are organisms, cells, or electronic components—link their repeating behavior for a common purpose, researchers have turned to the mathematics of oscillators. By using appropriate mathematical models, they can learn how communal rhythms may arise out of randomness and how such rhythms may break down into chaos.

Keeping Time

Late in the winter of 1665, an ailing Christiaan Huygens (1629–1695) was confined to his room for a few days. The Dutch physicist whiled away the hours of his confinement by closely observing and pondering the remarkable behavior of two pendulum clocks he had recently constructed. He noticed that the pendulums of the two suspended clocks, hanging side by side from a common support, were swinging together. When one pendulum swung to the left, the other went to the right. The pendulums remained precisely in opposite phase for as long as he cared to watch.

His curiosity piqued, Huygens began to experiment. He deliberately disturbed the motion of one pendulum so that it no longer mirrored the other's movements. Within half an hour, the two pendulums were back in opposite-phase motion. Huygens suspected that the clocks were somehow influencing each other, perhaps through air currents or vibrations of their common support. To test this notion, he moved the clocks to opposite sides of the room. They gradually fell out of step, with one clock losing five seconds a day in relation to the other. The two pendulums no longer swung at exactly the same frequency or in opposite phase.

Conversely, the clocks kept precisely the same time when placed side by side. In general terms, what Huygens had serendipitously discovered was a phenomenon that came to be known as the mutual synchronization of coupled oscillators. An oscillator is any system that displays repeated behavior. A pendulum or a child's swing, for example, swings back and forth at a constant rate. The steady vibrations of a sliver of quartz regulate the precisely timed ticks of a modern wristwatch. A firefly's biochemical cycle fuels a train of periodic flashes.

The oscillations generated by such a system can be characterized by its period (the time required for one complete cycle) and its waveform—how its amplitude, or strength, changes during each cycle. Many biological oscillators typically include damping mechanisms that keep oscillations from growing too large and sources of energy to keep them from becoming too small. A child on a swing experiences something similar: A swing has a natural frequency at which it moves to and fro; friction and air resistance keep the motion from lasting forever; and the child can put in energy by pumping to keep it going.

The behavior of a single oscillator is generally quite straightforward and predictable. When two or more oscillators are coupled in some way so that they influence each other, however, their joint behavior can become remarkably complicated. Often, it depends on the details of the strength and type of coupling and on whether the interactions involve only the nearest neighbors or all the oscillators of an enormous community.

Synchrony is perhaps the most familiar type of organized activity possible in a group of coupled oscillators. The fireflies of Asia represent a particularly spectacular case of synchrony arising out of interacting oscillators.

Night Lights

More than two thousand species of fireflies inhabit the temperate and tropical zones on Earth. Each species has a distinctive pattern of flashes, which serves as a sort of Morse-code identifier to help males and females find each other during a life span that barely lasts a week.

All of these beetles generate light in roughly the same way. A specialized light organ at the back of the insect's tail acts like a combustion chamber, where three ingredients manufactured by the firefly react with oxygen. A nerve impulse activates the chemical reaction, which generates a pulse of light. The light ranges in color from orange and yellow to green.

Among the Asian species that manage to synchronize their flashes, individual fireflies somehow react to sequences of sudden, distinct light pulses from their neighbors and adjust their rhythms accordingly. The entire performance occurs without the benefit of a maestro to bring the fireflies into rhythmic harmony.

Biologists have long wondered how these Asian fireflies manage to synchronize their flashes, and a few researchers have performed experiments to determine the conditions necessary for such synchronization to occur. In recent years, mathematicians have also entered the picture, developing a variety of mathematical models to elucidate the crucial features that may underlie coordinated activity.

In one approach, based on the simplifying assumption that fireflies react only to the sharply defined flashes of their neighbors, the mathematicians Steven H. Strogatz of Cornell University and Renato E. Mirollo of Boston College created a rudimentary oscillator model demonstrating that oscillators started at different times will always become synchronized. Their work was inspired by an earlier study aimed at shedding light on how the cells of a heart's natural pacemaker coordinate their individual electrical rhythms to generate a normal heartbeat. Consisting of a small mass of specialized muscle fibers, the pacemaker sends out electrical waves to the rest of the heart, causing the heart's chambers to beat and pump blood.

That effort was the work of Charles S. Peskin, a mathematician at the Courant Institute of New York University, who had become interested in the application of mathematics to biology, particularly the mechanics of how a heart pumps blood. In 1975, to find out what makes the heart tick, Peskin proposed a highly schematic model of how a mass of about ten thousand cells can beat together.

He started with equations representing a large number of identical oscillators, each one strongly coupled to all the others. To mimic the firing of a pacemaker cell, an electric current flowing into an oscillator would steadily increase the oscillator's voltage until it reached a critical value. At the threshold, the oscillator would discharge and the voltage would drop to zero. The voltage would then begin to rise again, restarting the cycle.

One oscillator could affect the others only when it fired, kicking up their voltage by a fixed amount. Thus, oscillators that were close to threshold would be pushed over the top, and others would move a small step nearer to firing. Peskin conjectured that a group of such oscillators would eventually become synchronized. But he ran into technical roadblocks when he tried to solve the equations that represented his model. He ended up examining the simplest possible case: two identical oscillators—representing two heart cells—that influence each other via their mutual signals. Peskin then proved that two such oscillators would eventually get in sync.

Whether the same thing would happen for a large number of oscillators, however, wasn't settled. Peskin suggested that it would. Nonetheless, with huge numbers of interacting units and conflicting signals, it's quite possible that something very complicated could happen. For example, the system could remain in a random state and never settle down. Or it could break up into distinct groups, with some individuals within a group becoming synchronized but the groups staying out of step with one another.

In 1989 Strogatz, who was then at MIT, came across a reference to Peskin's work in a book on biological oscillators by Arthur T. Winfree of the University of Arizona. Intrigued, he looked up Peskin's original paper on the subject and wrote a computer program to simulate the behavior of Peskin's model for not just two but any number of identical oscillators started at random times. He found that the system always ended up firing in unison.

The simulations showed that each oscillator initially receives many conflicting signals. As the system evolves, oscillators begin to clump together in groups that fire at the same time. As these groups get bigger, they produce larger collective pulses, which tend to bring other out-of-sync oscillators to the brink of threshold even faster. Large groups grow at the expense of smaller ones. Ultimately, only one huge group remains, and the entire population is synchronized. The key requirement for synchrony is that once a group of oscillators has locked together it stays together from then on.

With Mirollo, Strogatz tried many different initial conditions, and the system always ended up synchronizing. The results gave the mathematicians confidence that Peskin's original conjecture could be proved, establishing it as a mathematical theorem. Relying heavily on geometric arguments, they eventually showed rigorously that regardless of the initial conditions, oscillators operating at the same frequency but starting at different times will always become synchronized.

The simplified mathematical model developed by Strogatz and Mirollo was general enough to suggest possible application in a wide variety of systems in which oscillators interact via pulses, including the fireflies of Southeast Asia. For fireflies, the oscillator is the chemical cycle used to create a flash of light. The voltage of the theoretical model would then correspond to the beetle's readiness to flash. Every time one firefly sees another's light, the signal would serve as a stimulus that forces the firefly to fire a little earlier than it normally would in its cycle.

Strogatz and Mirollo's work has illuminated mechanisms that potentially lead to synchronization. For their model to work, they had to assume, for instance, that an oscillator leaks some electric charge as it gets closer and closer to threshold. Thus, the increase in voltage slows a bit as the oscillator gets charged up before firing. Real biological membranes appear to have a similar characteristic, in the form of channels that allow electric current to leak out of a cell. This leakiness plays roughly the same role that friction does in mechanics or resistance in electricity.

However, the equations used by Strogatz and Mirollo incorporated a number of simplifications that clouded their applicability to a swarm of real fireflies. For example, not all fireflies of a given species flash at precisely the same rate, and it's highly unlikely that every firefly sees the flash of every other firefly.

At the same time, experiments involving different species of fireflies in Southeast Asia demonstrate that each species tends to have a characteristic flashing frequency despite small differences among individuals in a group. Moreover, fireflies of a particular species have a fairly narrow frequency range, and they don't pay attention to anything flashing at a rate outside that range.

Entomologists have observed overlapping swarms of two different firefly species flashing independently at two synchronous rates. They have also noted that although firefly flashes ordinarily have fairly uniform intensities, durations, and delays, no single swarm ever achieves perfect synchrony. In fact, observers have noticed waves of flashing among fireflies, especially when the congregation is spread out over a large tree or in a string of bushes along a riverbank.

The Strogatz-Mirollo firefly model has additional deficiencies in accounting for firefly behavior. For instance, the model postulates that pulses and responses are instantaneous and that sensed pulses always advance an oscillator toward the threshold. However, firefly pulses clearly have a finite duration, and in some species of fireflies such signals can either advance firing or delay it.

Over the years, researchers such as Arthur Winfree and G. Bard Ermentrout, a mathematician at the University of Pittsburgh, have demonstrated that there are many different ways in which oscillators can become synchronized. The way fireflies synchronize their signals may have little in common with the way heart cells coordinate their beats or with the mechanisms underlying other types of communal activity. Different systems often operate in different ways.

Ermentrout, in particular, has developed models far more realistic and applicable to fireflies than that of Strogatz and Mirollo. One important feature that he includes in his mathematics is the ability of a firefly to alter the natural frequency of its flashes. He has shown, for example, that under various conditions, a set of oscillators connected in a ring can either synchronize or flash in sequence to create a wave of activity that travels around the ring.

Such mathematical models help highlight the importance of oscillations in biological systems and illuminate the factors that may lead to different types of synchrony, various traveling waves, or even completely disordered behavior. "The central question concerns how large numbers of oscillators get together to do useful things for an organism," Ermentrout says. The relevant mathematics is quite complicated and the resulting behavior remarkably diverse.

The mathematical analysis of mutual synchronization represents a challenging problem. Moreover, the synchronization demonstrated by fireflies—whether in the wild or idealized in a mathematical model—

isn't the only type of coordinated, symmetric behavior possible in coupled oscillators. Such alternative patterns of coordination have a bearing on the way in which a variety of animals move.

Doing the Locomotion

A visit to the zoo, especially one in which animals have plenty of room to exercise, can provide a vivid lesson in the different ways in which animals move, from the bounds of a kangaroo and the graceful leaps of a gazelle to the rocking gait of a cockroach and the slithers of a snake. It's possible to model such movements as networks of coupled oscillators. Consider the case of two identical oscillators. Such a linked pair can be either in step or out of step (in opposite phase). Two-legged animals show the same kind of synchrony and antisynchrony in their movements.

For example, when a kangaroo hops across the Australian outback, its powerful hind legs oscillate together, striking the ground at the same time with every hop. In contrast, a human running after the kangaroo plants one leg, then the other leg half a cycle later, with each foot hitting the ground alternately.

Larger groups of oscillators permit a wider range of movements, as shown in research by the mathematicians Ian Stewart of the University of Warwick and Martin Golubitsky of the University of Houston. Three oscillators joined in a triangular ring, for example, display four basic patterns.

Imagine a three-legged creature with sufficient agility and versatility to demonstrate the movements. It can hop by pushing its three legs off the ground at the same time, then landing on the three legs. It can scurry by having one leg at a time on the ground, with each leg taking a turn in succession. It can walk by planting first one leg, then a second, on the ground alternately, while its third leg hits the ground at the same rate as the others but at different times. It can also walk with its third leg striking the ground twice for every time either of its other two legs hits the pavement.

The discovery of the curious double-time oscillations in the fourth pattern initially surprised Stewart and Golubitsky. They soon realized, however, that such a pattern can occur in real life—in the shape of a person walking slowly while using a walking stick: right leg, stick, left leg, stick, right leg, stick, left leg, stick, and so on. In a sense, the faster oscillator is driven by the combined effect of the other two. The remaining three-oscillator patterns can be found among movements of quadrupeds missing a leg.

Four-legged animals can display many different gaits. A rabbit hops by moving its front legs together, then its hind legs. A giraffe also pairs its legs, but in this case it's the front and rear legs on each side that move together. The legs of a trotting horse are diagonally opposed in synchrony. Young gazelles sometimes make a movement called a pronk, a four-legged leap in which all four legs move together. Each of these movements can be represented mathematically by a system of four oscillators moving in different combinations and phases.

Some animals, like the elephant, have only one type of rhythm. In the elephant's case, that movement is a walk in which the animal lifts each foot in turn. To travel faster, an elephant simply makes the same movements more quickly. In contrast, a horse has a distinct gait at different speeds. At slow speeds, a horse walks; at moderate speeds, it trots; and at high speeds, it gallops.

Working with the biomedical engineer James J. Collins of Boston University, Stewart has also investigated the walking movements of insects, which must coordinate six legs to get anywhere. A cockroach, for example, has a "tripod" gait, with triangles of legs moving in synchrony: Front left, back left, and middle right legs lift in the first half of the cycle, and the other three legs lift in the other half.

The tripod gait is the most common type of movement among insects, and it occurs at moderate and high speeds. Such a gait represents a very stable pattern in a closed, hexagonal ring of six oscillators. At low speeds, insects often adopt a pattern in which a wave of leg movements sweeps from the back of the animal to the front, first on one side, then the other. Adjacent legs on one side of the insect move so that they are one-sixth of a cycle apart, and the two legs on opposite sides of each segment of the insect are half a cycle apart.

Other types of movements are possible. Cockroaches, for instance, have several different kinds of steps, which can be used to great advantage when they're scurrying across a kitchen floor or evading a broom. In 1991, the biologists Robert Full and Michael Tu of the University of California at Berkeley discovered that the American cockroach actually switches to a four-legged or even a two-legged gait when it needs to run quickly.

Whether in cockroaches or horses, the gaits observed among animals have an uncanny resemblance to the patterns of coordinated, symmetric movements that mathematicians can work out for various networks of oscillators. Such mathematical models serve not only to elucidate the mechanics of animal movement but also to guide the design and control of multilegged robots, especially for navigating over rough terrain.

The rhythms and patterns of animal gaits are apparently determined by a network of nerve cells. Such neural networks are known as central pattern generators. Biologists postulate that this neural circuitry generates the rhythmic electrical signals that trigger movement in animals. For instance, in principle it would take only a pair of coupled neural oscillators to drive the muscles that control the legs of a walking human or any other biped.

The mathematics of networked oscillators suggests that a single central pattern generator may be sufficient to control not only the movements of an animal such as an elephant, which has only one type of walking rhythm, but also the varied gaits of animals like horses. In theory, the different patterns of leg movement required for each gait could arise from different signals driving the same central pattern generator.

In real life, depending on the animal, a neural oscillator could be as simple as a single nerve cell or as complex as a jungle of interconnected neurons. Determining cause and effect in neural systems responsible for walking, swimming, chewing, and other rhythmic activities involves untangling the threads by which brain and nerve take charge of life.

Swimming with Lampreys

With gaping, round mouths ringed by sharp, horny teeth, lampreys are among the most primitive of aquatic vertebrates. Lacking bones, these underwater vampires slither along with a rapid, undulating motion created by rhythmically flexing muscles to generate waves that travel from head to tail down their eel-like bodies.

The bursts of muscle activity, which bend the lamprey's spine first to one side, then the other, result from coordinated electrical pulses governed by a network of nerve cells that constitutes a central pattern generator. Such a neural network faithfully executes a particular action over and over again without the need for conscious effort. Similar mechanisms underlie propulsion in a variety of swimming animals, from water snakes to many fish.

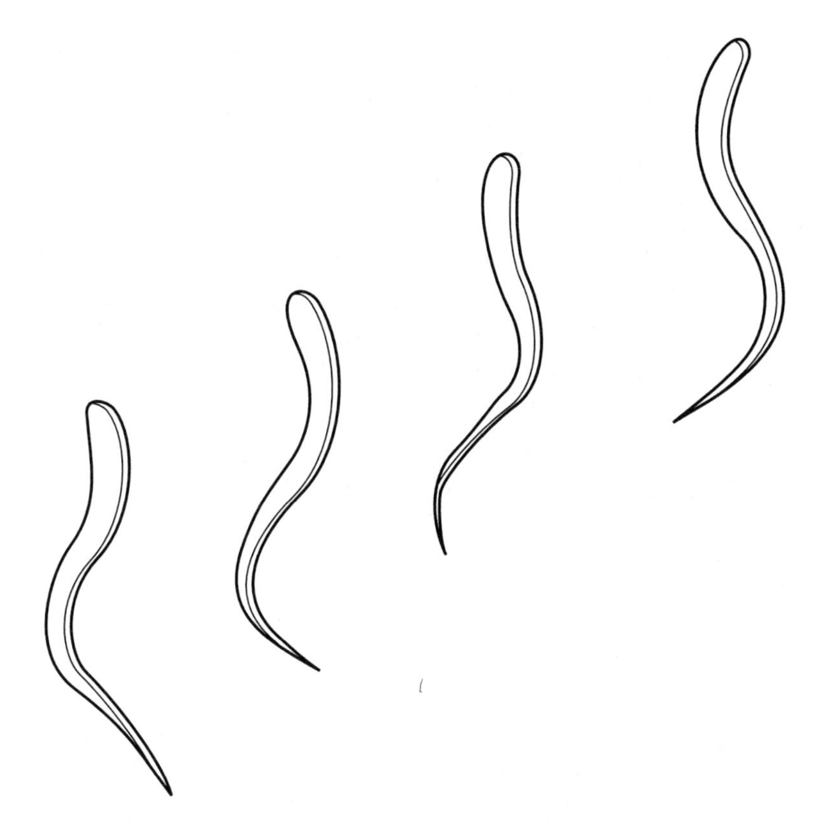

A lamprey swims forward by generating characteristic S-shaped waves that ripple its elongated body.

The human body also performs numerous everyday miracles of motion that require no conscious effort. When a person walks, neural control circuits rooted in the brain stem, which surrounds the uppermost segment of the spinal cord, signal a step and the correct set of pulses automatically races down the spinal cord to activate the appropriate muscles. The right joints move to just the right degree at just the right time.

Indeed, whether the action is walking, swimming, breathing, or chewing, nerve cells must signal one another in a kind of rhythmic electrochemical chatter that eventually translates into muscular action. One key to deciphering communication in such a network is to determine how the network and the signals it carries are organized and to tease out the signals that matter most in deciding a certain behavior.

The trouble is that neural networks are typically very complicated. A human, for instance, has about a trillion neurons distributed throughout the body. It turns out, however, that movements like walking originate in an animal's spinal cord, not in the brain. That realization came early in the twentieth century when neurophysiologists observed that mammals with a severed spinal cord could still generate alternating leg movements even though the connection to the brain had been cut.

Additional experiments showed that the brain's main role in controlling movement is to inhibit the signals automatically generated in the spinal cord—until the movements are required for some purpose. If disease or injury affects that part of the brain, the victim often displays a variety of uncontrollable, involuntary twitches or spasms. Curiously, whereas the spinal cord itself generates the intricate patterns of impulses necessary for walking and running, the brain stem houses the specific circuits that regulate breathing, swallowing, and chewing.

Because the lamprey has only about a thousand nerve cells in each segment of its spinal cord, this animal serves as a suitable starting point for untangling how a central pattern generator produces movement. Moreover, stripped of brain and muscle, a lamprey's bare spinal cord, when immersed in a saline solution spiked with an appropriate amino acid, generates the same traveling waves of electrical activity that is seen in the intact animal. With such a convenient, relatively simple system, both biologists and mathematicians improve their chances of solving the puzzle of its smoothly coordinated swimming motion.

From a mathematical point of view, a lamprey's spinal cord is like a chain of coupled oscillators, in which each mathematical oscillator corresponds roughly to a group of cells along the lamprey's spinal cord. The trick is to identify what sequence of electrical signals along the chain produces the same kind of wave that passes down the length of a lamprey's body as it swims.

The model developed several years ago by the mathematicians Nancy Kopell of Boston University, Pittsburgh's Bard Ermentrout, and others takes the form of a neural network made up of a chain of subnetworks, each link of which can produce its own rhythm. Describing such networks in the form of equations, the researchers could get a wave to travel along their mathematical chain, just as it does along a lamprey's body.

Interestingly, the mathematicians were able to identify types of behavior in their lamprey model that biologists had apparently overlooked in their laboratory studies. For example, the model predicted that, in order to get waves, the connections from tail to head determine both the direction of the wave and the wavelength of the wave, in spite of the fact that the wave itself travels from head to tail.

That result ran counter to a notion intuitively held by biologists: that because the wave associated with swimming travels from head to tail, head-to-tail neural connections would probably be stronger than tail-to-head connections. To test the mathematicians' prediction, experimental biologists searched for a similar effect in spinal cords extracted from lampreys. To everyone's astonishment, the mathematical prediction proved to be correct. With one end free to move and the other end periodically wiggled by a small motor, the cord's electrical activity exhibited the same asymmetry present in the mathematical model.

The most important mathematical result, however, was the demonstration that key aspects of the neural activity leading to undulatory motion don't depend on the specifics of the participating neurons or the chain of oscillators. Changing various characteristics of the oscillators does little to disturb the waves that propagate along the chain. What matters most in this particular case is how the signal travels from unit to unit, not the nature of the units themselves. Nonetheless, there are many other situations in which the details are critical, and one can use mathematics to discover which details matter most.

The mathematics used by Kopell and Ermentrout has helped researchers pinpoint, in a mountain of marginally relevant detail, important factors establishing a lamprey's swimming rhythm. "If you know some of the properties of the subnetworks making up the larger network, and you know something about how they communicate, then you can still derive robust explanations," Kopell says. "Based on our findings, we made predictions that have held up in the lab—much to our surprise and delight."

More recent studies have now allowed biologists to reconstruct the specialized neural circuitry that governs how lampreys swim. Using this information, researchers have developed computer simulations that not only mimic the lamprey's neural activity but also incorporate other components, from the muscle fibers controlling different segments to the flow characteristics of the surrounding water. The resulting virtual lamprey is an excellent swimmer—at least on the computer screen.

Kopell's efforts to understand how behavior at one level affects that at another often involves the use of mathematics, for example, when she explores how the properties of individual cells affect network activity. Neurons, for instance, are extremely varied in size, structure, and composition, ranging from long, thin, spindly strands to dense clumps with numerous branches. In many situations, the structural details and

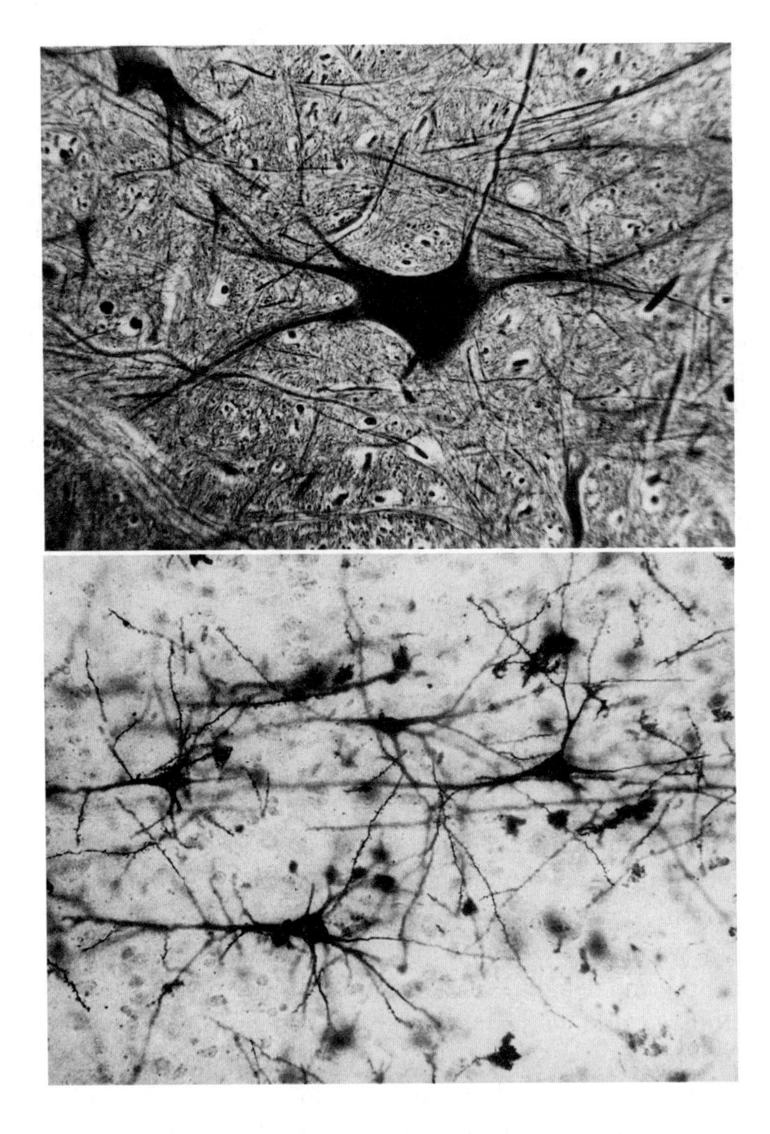

Neurons are extremely varied in size, structure, and composition, ranging from long, spindly strands to dense, heavily branched clumps. Above is a motor neuron cell from the spinal column and below are larger neuron cells from the cortex in the brain.

the chemistry of how the cells operate strongly affect how they speak to one another.

In recent years, Kopell's focus has been on how cellular properties affect oscillations in the brain, such as the rhythms of human sleep. "Researchers are finding oscillations all over the nervous system," Kopell notes. "What's not understood is what the oscillations are doing. The kinds of questions we're asking about rhythms and network behaviors can be addressed only through modeling. Classical biological techniques of electrophysiology or anatomy can't address many systems-level questions about how the parts work together, just as knowledge about the components of a computer doesn't, by itself, explain how the computer works."

The human central nervous system is enormously, almost overwhelmingly complex. It orchestrates sleep, movement, vision, thought, daydream, and much more. A great deal of this activity is rhythmic; some of it may be chaotic. "As always, we are left with a feeling of awe for the amazing complexity found in nature," says Christof Koch, who is a professor in the computation and neural-systems program at the California Institute of Technology. "Loops within loops across many temporal and spatial scales. And one has the distinct feeling that we have not yet revealed every layer of the onion."

Immersed in such unavoidable complexity, one turns to simple organisms and simple models to glean insights that may provide a glimmer of understanding of our own rhythms and natural clocks.

Rhythms of Life

A living cell is much more than a little bag of chemicals. It's continually in action: signaling, copying, reproducing, tearing apart, and rebuilding. As a cell changes over time, its activities often have characteristic rhythms. Those rhythms, in turn, reflect fundamental natural cycles, from the recurring process of cell division to the twenty-four-hour sun-driven clock that marks each day.

Biological clocks, set according to the sun or to some other cycle, play an important role in the lives of many organisms. They control the timing of leaf movements in plants, running activity in hamsters, the predations of mosquitoes, and, in humans, the rise and fall of body temperature, the setting of sleeping and waking cycles, and even the ups and downs of mental alertness. In each case, we can observe the cycle, and we try to deduce the pacemaker—the underlying timing mechanism.

One of the more curious examples of such a cycle in a simple organism involves the luminescence of a type of algae known as *Gonyaulax*. Along the Pacific coast of the United States, a nighttime swimmer can readily disturb a patch of these single-celled organisms into emitting light and making the water glow.

By growing cultures of *Gonyaulax* in the laboratory, researchers have discovered that the algae stop producing light at dawn, and they begin luminescing again in the evening. Shaking a vessel containing thousands of cultured cells during the night causes them to emit bright flashes of light that last less than a tenth of a second each.

If the organisms are illuminated continuously with a dim, steady light so they are no longer exposed to a normal day-night cycle, something interesting happens. Shaking the vessel at different times reveals that the maximum output continues to occur each night at about 1 A.M. and decreases to a minimum twelve hours later. The algae retain their communal cycle, maintaining the rhythm for weeks.

Eventually the periodicity apparently dies out. It turns out, however, that individuals continue to oscillate with a roughly twenty-three-hour period. But because they have slightly different natural frequencies, they get out of step. When one is ready to flash when disturbed, another may still be slumbering. Without the sun to keep their clocks precisely in step, the response of the organisms to the time of day drifts out of synchrony.

In general, the behavior of communities of oscillators whose members have clocks with different natural frequencies depends on the strength of the coupling that takes place among them. If their interactions are too weak, the oscillators are unable to achieve synchrony. Even when they start together they gradually drift out of phase. In other communities, coupling may be strong enough to overcome the differences in individual frequency that inevitably occur—just as Huygens's pendulum clocks had to be linked in some way through a common support or air currents in order to affect each other when they were close together. Such groups can maintain synchrony for long periods of time.

The mathematician Norbert Wiener (1894–1964) was among the people who were impressed by the ubiquity of communities of coupled oscillators in nature. In the 1950s, he was already well known for his contributions to the development of computers and as the founder of the field of cybernetics, which concerns the analysis of control and communications systems in machines and in living organisms. By that time, Wiener had become interested in the electrical signals generated by the human brain.

Collaborating with medical colleagues, he made recordings of the brain signals of subjects who were awake but resting with their eyes closed. Under those conditions, the resulting electroencephalograms showed considerable activity at frequencies of around ten cycles per second, a characteristic signal now known as the alpha rhythm. Wiener also observed that the spectrum of frequencies had a distinctive shape, as the figure below demonstrates.

Recording his impressions, Wiener wrote: "There are two things that are striking here: One is the very narrow line at the center of the spectrum, and the other is that this line arises from a dip." To explain the spectrum, he suggested that the brain incorporates a population of neurons operating at frequencies close to ten cycles per second. To generate such a sharp signal, he hypothesized that these neural oscillators interact by pulling on each other's frequencies. If an oscillator is ahead of the group, the group tends to slow it down. If it is behind, the group tends to speed it up.

As a result, most of the oscillators end up at the same frequency, creating the large spike in the signal. The dips on either side of the central peak signify the absence of oscillators whose frequencies were shifted to the communal standard. The peaks on either side of the main signal represent outliers—renegade oscillators whose frequencies are too different to be pulled into the group.

The phenomenon of phase-locking, in which fast oscillators get slowed down and slow ones are speeded up to lock the entire ensemble into a compromise rate, allows a community to be a more precise timekeeper than would be possible for any individual member of the community. Thus, nonidentical oscillators can interact to beat with a single rhythm.

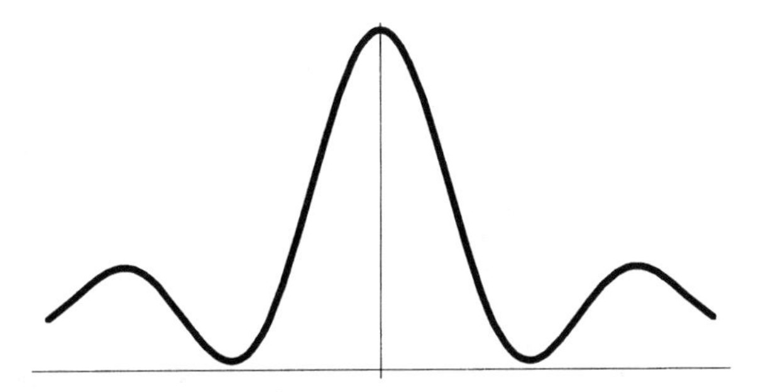

Measurements of electrical brain activity show that people who are awake but resting with their eyes closed produce a characteristic brain-wave rhythm. The frequency spectrum of these so-called alpha waves has a distinctive shape, showing a peak near ten cycles per second and little activity at frequencies close to the peak.

In the late 1960s Arthur Winfree, who was then a graduate student at Princeton University, provided a mathematical framework for dealing with the mutual synchronization of coupled oscillators. Much of the experimental work in testing his ideas involved an ingeniously constructed array of seventy-one flickering, electrically coupled neon lamps, which he dubbed the firefly machine. There were seventy-one lamps, he notes, because out of the hundred that he started with, twenty-nine wandered too far out of the proper frequency range during the initial "breaking in" period to be useful in his experiments.

"The purpose of building this machine was just to 'look and see what would happen,' on a hunch that groups of oscillators might synchronize together in fleeting alliances," Winfree says. "One hope was that by plotting the output of this population of interacting oscillators in the same format as biologists use to plot activity rhythms of multicellular animals, enough resemblances might be noticed to suggest some interpretation of the tantalizingly complex biological records."

Winfree found that with no coupling, all the neon oscillators ran at their own frequencies, losing their initial synchrony. Only the most fleeting, chance alliances occurred among the oscillators. Curiously, simple electrical links between the oscillators also proved insufficient to guarantee mutual synchrony of the group even though such links would lock oscillators together in pairs.

A stronger form of electrical coupling, however, did induce synchrony, which happened abruptly among the neon oscillators once a certain threshold was reached. One could also observe small groups of neon oscillators temporarily escape entrainment, only to reunite with the majority at some later time.

Combining his firefly-machine experiments with computer simulations and mathematical analysis, Winfree was able to develop equations that seemed to capture key aspects of rhythmic biological behavior. His theoretical approach still guides investigations of large systems of coupled oscillators in biology—whether they be algal cells, fireflies, or neurons.

In his studies of biological oscillators, Winfree found it particularly useful to assume that the oscillators are self-sustained, meaning that a biological oscillator typically has a stable, innate oscillation rate. Jolted by a shout, for instance, your heartbeat may start racing. But if your heart is healthy, the heartbeat will quickly settle back to its normal rhythm.

To obtain more readily solvable equations, Winfree also introduced several simplifications, including the assumption that each oscillator is influenced by the collective rhythm produced by all the others. That's like saying that each firefly responds to the combined flash of all the other fireflies of the swarm rather than to the flash of any individual firefly.

Though Winfree's model was only minimally realistic in terms of biology, it allowed him to work out a possible scenario for synchronization. He suggested that the onset of synchronization is like a change of state, such as the freezing of a liquid. Initially, the oscillators behave incoherently, with each one running at its own pace. Gradually, however, fast oscillators slow down and slow ones speed up. When the spread of frequencies among the oscillators decreases to a level below a certain critical value, the system spontaneously "freezes," and the entire ensemble locks into a single, compromise rate.

Winfree's work was just the beginning. Many mathematicians, biologists, and others have since extended and refined it. These efforts, however, have barely encompassed the incredible variety of dynamics possible with coupled oscillators, whether they appear in pairs and threesomes or in enormous groups.

Keeping Pace

The mathematics of coupled oscillators has helped elucidate the mechanisms that govern such rhythmic behavior as walking, swimming, and breathing, not only for understanding biological systems but also for designing robots. At the same time, applications of mathematics to biology suggest new mathematical phenomena worthy of investigation.

Mathematicians can apply two types of strategies to such efforts. The first is to study in detail what happens with just two or three oscillators. Recent work shows that even two coupled oscillators can produce dynamics of bewildering complexity. A second approach involves investigating enormously large systems of oscillators. The trick is to forget the details and focus on the collective activity. However, nearly all existing theoretical work involves rather drastic simplifications and unrealistic assumptions. Overcoming any of the constraints could lead to new insights and suggest new problems worthy of mathematical investigation.

In all this mathematical modeling, however, there remains a constant tension between what we have learned to solve and what we need to solve in order to capture the crucial and relevant aspects of a biological system at work, whether it's a lobster rhythmically digesting its food, a firefly flashing its signature greeting, or a baby taking its first steps. Sometimes we have to make do with a simplified model, though we know it doesn't provide the complete answer. With luck, the simple model may highlight the features that really count, and we can learn from what the equations predict.

Mathematicians trying to understand biological oscillators face difficult mathematical questions. "It would be very desirable to start building in a little more reality," Strogatz says. But, as so often happens in mathematics, one problem may turn out to be relatively easy to solve, while "everything else in every direction around you is hard," he remarks.

Biologists have much to learn, too. Firefly behavior alone is remarkably diverse, complex, and not yet fully understood. Investigators have learned a great deal about the synchronous rhythmic flashing of fireflies in the last fifty years, noted the firefly specialist John Buck in a 1988 review paper on their behavior. "At the same time its mysteries have multiplied," he continued. "Each step of physiological elucidation has revealed new black boxes and each behavioral insight has left major puzzles yet unsolved."

Future progress in understanding biological oscillators may depend on greater cooperation between mathematicians and biologists. "Best of all," says Strogatz, "would be to collaborate with a biologist who actually measures things in fireflies against which to check the quantitative predictions of mathematical models."

Defining and characterizing the wonderful intricacy of nature—even just to capture the elusive ingredients of harmony among oscillators—represents a task of overwhelming complexity. Efforts to tame such complexity require cooperation among researchers and the development of communities that can work together and contribute diverse perspectives.

What of oscillators themselves, particularly those that generate the waves in the world around us—waves that carry information for us to decipher? Randomness, chance, and physical law play their part in confounding the signals that we try to interpret. The sounds of drums carry a disturbing message about what we can and cannot learn from vibrating structures and their characteristic frequencies.

5

Different Drums

Now I will do nothing but listen . . .
I hear all sounds running together, combined,
fused or following,
Sounds of the city and sounds out of the city, sounds
of the day and night. . . .

—*Walt Whitman (1819–1892)*, "Song of Myself"

The sounds of a drowsy summer afternoon weave through the shuttered house. From outside comes the faint drone of a distant lawn mower, the warbling and trilling of birds in a nearby shade tree, the shrill, pulsating sighs of a chorus of crickets and cicadas, the steady thump of a ball bounced on the pavement, and a child's jubilant shout and quick laugh. Inside, an air conditioner clicks into rumbling motion, the refrigerator emits an insistent whine, a telephone chirps briefly then subsides into silence, a computer hums as keys clack, and a chair creaks.

Tangled waves of sound wash over the listener. His ears capture the motion of air. An eardrum vibrates, rocked back and forth in tune with the aerial ripples. Three delicate bones—a hammer, an anvil, and a stirrup—transmit the movements to the inner ear, where millions of microscopic hairs convert the vibrations into electrical nerve signals to be interpreted by the brain.

Eyes shut, the listener sorts through the signals, ignoring some, focusing on others, picking out and identifying individual sounds in a mad mixture of acoustic waves. It's a remarkable process, one that we normally can't evade. In *The Tuning of the World,* the musician and composer R. Murray Schafer noted: "The sense of hearing cannot be closed off at will. There are no earlids. When we go to sleep, our perception of sound is the last door to close and it is also the first to open when we awaken."

Our auditory system is adept at analyzing and interpreting sound waves. We have also fashioned a wide variety of instruments for detecting and unsnarling other kinds of waves, from the seismic tremors that signal massive shifts in Earth's drifting and crinkled crust to the X rays that penetrate our bodies to reveal the bones and organs inside.

The business of analyzing waves to glean useful information brings up the question of the relation between a vibrating system—whether it is a violin string, a contorting molecule, or Earth's crust—and its characteristic oscillations. How likely is it that different systems can generate identical waves that could fool us into incorrect interpretations?

Recent discoveries concerning novel solutions of a mathematical expression known as the wave equation suggest that such ambiguities may very well arise in physical situations that involve waves. These findings have important implications, because much of what we know of the natural world comes by means of indirect measurement rather than from direct observation. Astrophysicists determine the composition of the sun not by visiting it but by detecting the wavelengths of light emitted by the sun's excited atoms. Geophysicists construct a picture of Earth's interior not by journeying to its center, but by conducting studies of seismic waves. Physicians often locate ailments not by cutting open the body but by studying X-ray images or CAT scans.

How much can we learn about a puddle from the ripples that ruffle its surface? More generally, what can we tell about objects from measurements of their characteristic frequencies? To what extent can we reconstruct the medium from the effect? These are subtle, mathematical questions that lie at the heart of efforts to understand the meaning of the sound waves and other oscillations in the world around us—pockets of order and purpose amid the noisiness of the random face of nature.

Plane Sounds

The trumpets, flutes, and drums of a marching band are readily distinguishable by our ears even before we see the instruments. Moreover, different types of drums have distinctive sounds, ranging from the thundering boom of a bass drum to the rat-tat-tat of a snare drum.

Drums don't just produce noise—random vibrations; each type of drum has a particular mechanism for generating pressure waves that spread through the air. An orchestral snare drum, for example, has two heads—sheets of mylar or another elastic material stretched over the open ends of a squat cylinder. Strands of wire or gut, called snares, extend across the lower head. When a player strikes the upper head, those vibrations are transmitted to the snare head, and the joint movements of the heads and snares generate a sharp rattle. Moving the snares away from the lower head creates a very different kind of sound.

Though drums can be constructed entirely of wood, stone, or metal, the most familiar types consist of a membrane of animal skin or a synthetic material tautly stretched across some type of air enclosure. When the drum is struck, the hide or membrane vibrates, while its rigid frame remains fixed. In general, the characteristic sound generated by a given drum depends on its size and shape, on the thickness, uniformity, composition, and tension of its sound-producing head, and on the way it is beaten. What makes these sounds so identifiable is the fact that each drum vibrates at characteristic frequencies. The set, or spectrum, of pure tones produced by a vibrating membrane stretched across a frame gives a drum's sound its particular color.

One of the simplest examples of a vibrating acoustic system is a string with fixed endpoints—like the strings of a violin, guitar, or piano. Such a string is a sort of one-dimensional drum. When plucked or hammered, it moves in a pattern described by a mathematical expression known as a wave equation. This equation incorporates the notion that the farther the string shifts from its resting position, the more strongly it bends and the greater the restoring force exerted on it. Solving the equation means finding a mathematical expression that defines the shape of the string, or the position of any point along the string, at any time during the course of a vibration.

The string's motion is an oscillation. The restoring force pulls the plucked string back toward its original position, but the string has so much momentum that it overshoots. As the string moves away from its rest position, the restoring force increases to slow it down and bring it back. The restoring force weakens during the return, and the string again overshoots.

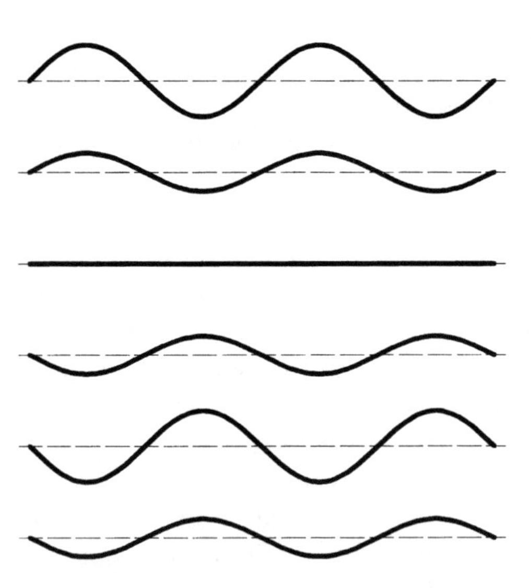

As a stretched string fixed at its two ends oscillates back and forth, it maintains a characteristic waveform. Only the amplitude of the vibration changes, as shown in this sequence of snapshots of a string's movements over a little more than half a cycle. The dashed line represents the string's rest position.

Among the solutions of the wave equation for a vibrating string with fixed ends are a set of equations that corresponds to standing waves along the string. These particular motions are called normal modes, and each one has a specific frequency, corresponding to the overtone series that is familiar to musicians. The lowest frequency is the fundamental, which establishes the string's pitch. The other frequencies, or overtones, are whole-number multiples of the fundamental frequency.

We generate notes of different pitch by using strings of different length, tension, or density. A violin string that sounds a lower note, for example, is generally thicker and heavier than its higher-pitched counterpart. Tightening a string to increase its tension raises its pitch. The sound produced by a guitar or violin string, however, is generally a complex waveform that contains a mixture of several frequencies. What we hear can be thought of as the output from combining the fundamental with certain doses of various overtones. Because this spectrum of overtones can vary from one type of string to another, it's possible to distinguish the sounds of such strings. Indeed, any sound—whether it's the toot of a train whistle or the roar of a jet engine—can be broken down into a spectrum of pure tones, with different tones contributing different amounts to the full sound.

By solving a wave equation to determine the vibrations of a string, one can, in effect, deduce the vibrations from the shape of the vibrating body. Doing the same for a vibrating drumhead is a trickier proposition.

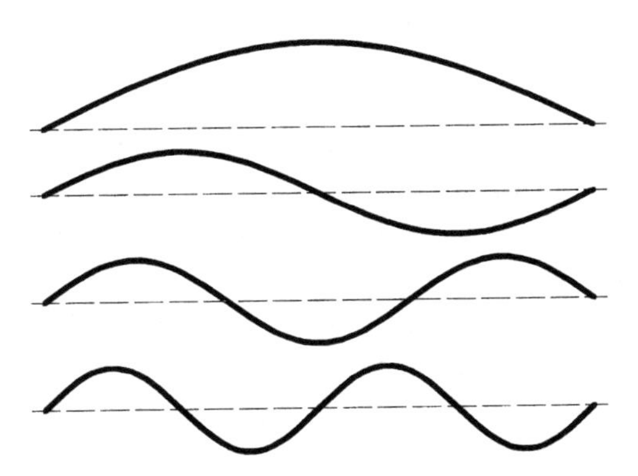

A stretched string vibrates at a fundamental frequency (top) and at whole-number multiples of that frequency (below), which constitute its overtones. Together, these vibrations are considered the string's normal modes.

Simplified and idealized to its mathematical essence, a drum is basically a flat, two-dimensional surface firmly fastened to a rigid rim. Only the interior vibrates, and this greatly restricts the surface's possible motions. Indeed, without such boundary conditions, a drum could make any sound. It would be like a flag flapping in the wind.

As in the case of a string, the vibrations, or normal modes, of a drum correspond to the solutions of a wave equation. This time, however, the equation's solutions are for a two-dimensional, or plane, surface, specifying how high or low each point within the boundary moves.

Over the last two centuries, physicists and mathematicians have developed ways of solving the wave equation to determine the normal modes of various drum surfaces. The shape of the boundary, whether it's a circle, an ellipse, a square, a polygon, or some other form, has a significant effect on both the frequency and the geometry of the surface's normal modes. For an ideal, circular membrane—one that is perfectly flexible and vibrating in a vacuum—the normal mode frequencies are given by a complicated formula. In this case, the different overtones are not whole-number multiples of a fundamental as they are in a vibrating string. Moreover, the vibration patterns are generally more complex than those for a string.

The fundamental, for example, corresponds to the entire drumhead repeatedly curving into a bowl, then into a dome. The next normal mode, at a frequency 1.594 times that of the fundamental, has the drumhead divided in half, with one half rising as the other dips, and vice versa. No motion occurs along the line separating the two regions. This line of no activity is known as a node.

Drumheads of different shapes have different normal modes, or sets of overtones. Finding formulas for calculating their frequencies and depicting the corresponding vibrational motions, however, has proved immensely difficult for all but such simple shapes as circles and squares or rectangles. In most cases, researchers must use computers to find approximate solutions of the relevant wave equation.

In principle, there appears to be no fundamental problem standing in the way of determining the normal modes, or characteristic vibrations, of a drumhead, no matter what its shape. It's also possible, however, to ask whether one can reverse the procedure—that is, whether one can definitively determine a drum's shape just from a knowledge of its normal modes. In other words, can one hear the shape of a drum?

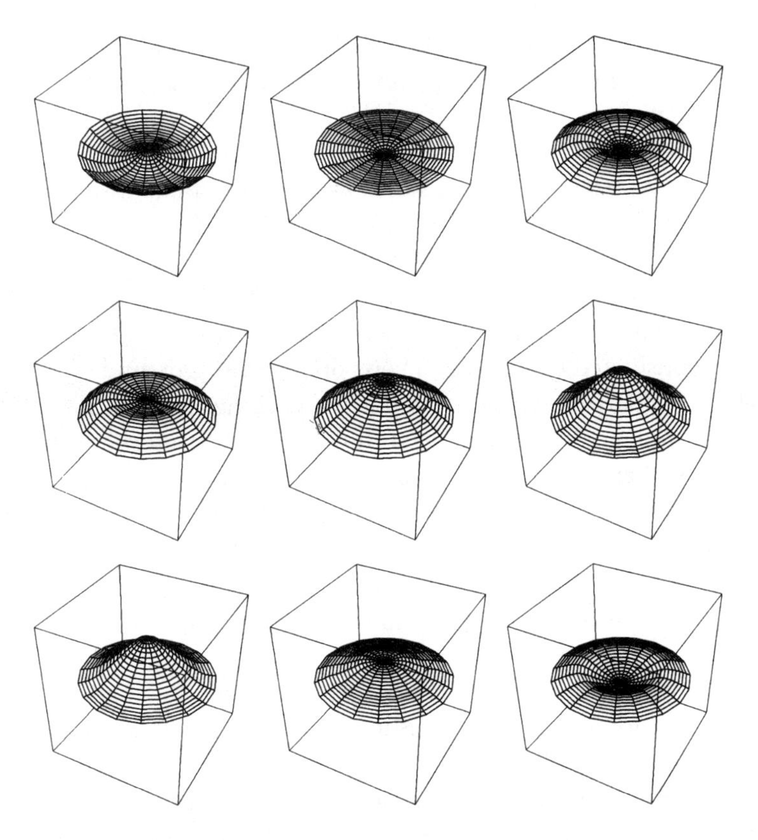

The normal-mode vibrations of a stretched membrane, or drumhead, have characteristic shapes. In the first normal mode, a circular drumhead oscillates between a dome shape and a bowl shape. The normal mode shown here has a more complicated geometry, as illustrated in these snapshots of its configuration at various times during one cycle.

Hearing Geometry

At the beginning of the twentieth century, physicists were deeply enmeshed in reworking their understanding of matter, space, and time. They had postulated the existence of electrons as fundamental particles of matter. They were beginning to understand electromagnetic waves—the carriers of visible light and radio signals—as the result of the accelerated motion of electrons. Albert Einstein (1879–1955) was just a few years away from formulating his special theory of relativity.

The Dutch theoretical physicist Hendrik A. Lorentz (1853–1928) was a major contributor to this revolution. His theory of electromagnetic radiation prepared the way for Einstein's ideas. In 1910, Lorentz presented a provocative lecture concerning the behavior of electromagnetic waves in a box—a step toward understanding how atoms emit light. He conjectured that a box's volume can be determined from the normal modes of the standing waves in the container, and he challenged mathematicians to provide a proof.

One of the academics attending Lorentz's lecture was David Hilbert, who is considered by many to be one of the most profound of modern mathematicians. There is a story that, on hearing Lorentz's challenge, Hilbert predicted that the conjecture would not be proved in his lifetime. Less than two years after his prediction, however, one of Hilbert's own students, Hermann Weyl (1885–1955), provided a proof that knowledge of the normal modes of vibration of an oscillator are sufficient for establishing its volume.

Mathematically, that was equivalent to proving, in two dimensions, that the overtones of a drum—abstractly, a membrane fixed along its perimeter—determine its area. Thus, by listening to a drum, it's possible to deduce its area. Simply put, the bigger the drum, the lower its tone, regardless of its shape.

In 1936, a Swedish mathematician went one step farther and showed that the length of the boundary of a membrane—a drumhead's perimeter—can also be determined from its overtones. Because the circle is the only shape that is exactly defined by just its area or its perimeter, it's easy to establish from its sound that a drum is circular.

Then, in the 1950s, the mathematician Mark Kac suggested, and others soon proved, that if a membrane has holes, the number of holes can also be calculated from the overtones. Hence, the area, perimeter, and connectivity (number of holes) of a membrane are audible, all leaving distinctive imprints in a drum's spectrum of characteristic frequencies.

These advances, however, still left open the more general question of whether knowledge of the normal modes allowed one to deduce a drumhead's precise shape unambiguously. In other words, given two membranes of equal area and perimeter, but different geometric shape, could they still have the same set of overtones, so that no one could tell them apart just from knowing the spectrum of normal modes?

In the 1960s, Kac focused attention on the matter in a major lecture, which was followed by the publication of a paper playfully titled

"Can One Hear the Shape of a Drum?" Because this title was far more evocative than usual for a mathematical treatise, the paper attracted a great deal of attention and proved remarkably influential, prompting a number of efforts to resolve the issue.

The question of whether knowledge of a drum's normal-mode vibrations is sufficient to allow one to infer its geometric shape is a particular example of what mathematicians call an inverse problem. The same idea underlies any technique in which an object is reconstructed out of measurements of certain observables, whether those objects are the layers of rock beneath Earth's surface or a cancerous tumor in a human body and whether the observables are seismic waves or X rays.

In fact, one *can* hear the shape of a string, which is a one-dimensional drum. The shape of the string is completely captured by its length, and it's not difficult to deduce this length from the value of the lowest frequency. It's plausible, then, to suppose that the sounds of a drum may encode enough information for specifying its shape uniquely. By the 1960s, however, evidence that this might not be the case had begun to accumulate.

Mathematicians aren't constrained by the three dimensions of ordinary space. They can formulate equations that work with any number of dimensions, define abstract spaces for these expressions, and calculate various characteristics of the geometric shapes that arise in rather strange, unimaginable domains. So, one can think about waves generated not only by one-dimensional strings, two-dimensional membranes, and three-dimensional boxes but also by eight-, ten-, or hundred-dimensional shapes known as manifolds. A manifold represents a specific type of shape, generally one that has no wrinkles, sharp folds, or cusps. Just as a small piece of the surface of a doughnut looks like a slightly warped section of a plane, sufficiently tiny pieces of a manifold look roughly like small fragments of ordinary space.

In 1964, John Milnor of the State University of New York at Stony Brook discovered a pair of geometrically distinct, sixteen-dimensional sound-alike drums, known technically as isospectral manifolds. He imagined that these objects were constructed in the same way that, in lower dimensions, it is possible to glue together the opposite sides of a flexible, rectangular sheet of rubber to create a doughnut, or torus. His two different, sixteen-dimensional doughnuts vibrated at exactly the same frequencies.

Other mathematicians found additional examples in different dimensions. For a long time, however, there appeared to be no pattern to their occurrence. These pairs of curious forms seemed to pop up randomly, scattered throughout the realm of multidimensional geometry. And there was no real clue to how such objects could be found in the two-dimensional world of drumheads.

One important breakthrough occurred in 1984 when the mathematician Toshikazu Sunada of Tohoku University introduced a method that made it possible to construct examples of isospectral manifolds almost at will and to determine their normal modes. Moreover, he provided a new, systematic approach for comparing sounds made by different shapes. Sunada's work spawned a mathematical cottage industry devoted to creating low-dimensional examples, including forms that one could cut out of paper and assemble with tape. These special shapes weren't flat, however. Curving around in three dimensions, they were more like exotic bells than drums.

Dennis DeTurck, a mathematician at the University of Pennsylvania, went so far as to use the calculated normal modes of some of these forms to synthesize a weird form of music. He created the sounds of different manifolds by putting together mixtures of their characteristic frequencies. Then, by connecting several, different-sized copies of the same manifold to a keyboard, he obtained a unique instrument on which he could play his "sphere and torus" music. Listeners attest to the fact that the sounds of the manifolds chosen by DeTurck are somewhat painful to the untrained, unsuspecting ear. Fortunately, perhaps, no pair of his examples produced the same terrible sound.

However, there are pairs of different bell-like shapes that do have identical normal modes, and it was one of these pairs that provided the crucial step that allowed the mathematicians Carolyn Gordon and her husband, David Webb, of Washington University, aided by a hint from Scott Wolpert of the University of Maryland, to settle Kac's question about drums once and for all.

Squashed Bells

The setting was a geometry conference held at Duke University in the spring of 1991. As Carolyn Gordon presented a lecture surveying progress in attacking Mark Kac's drum question, she held up paper models of a pair of three-dimensional shapes originally cooked up by the Swiss mathematician Peter Buser. These curious bells had the same normal modes, even though they had different geometries.

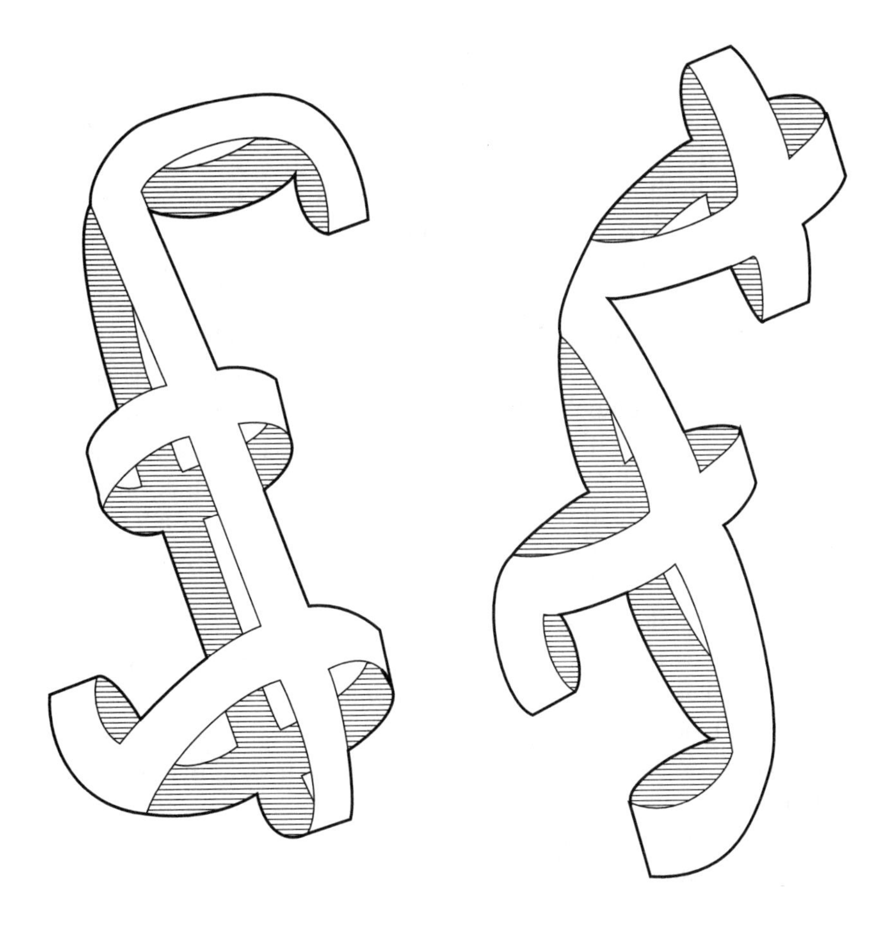

When these two soundalike "bells" ring, they generate the same set of normal modes.

At the conclusion of Gordon's lecture, Wolpert came up to ask her a simple question: What would happen if the two forms were flattened into planar objects? Would the resulting two-dimensional shapes also have identical normal modes? Wolpert suspected that they would.

In the end, Wolpert's hunch proved to be correct, but it took Gordon and Webb many months to come up with a proof. The shapes that resulted from flattening were simply too complicated to allow an exact calculation of their spectrum of sounds, and the mathematicians had to resort to more indirect means in order to accomplish their goal.

Initially, Gordon and Webb spent a week or so looking for a pair of shapes whose geometry was more amenable to calculation and manipulation than the pair they had in hand. They spent much of the time building intricate paper models of three-dimensional surfaces that would serve as candidates, but their investigations led nowhere. They finally returned to Buser's pair. The mathematicians ended up using techniques developed by the French mathematician Pierre Berard of the Fourier Institute, who had generalized Sunada's method to work with objects known as orbifolds. Orbifolds are like manifolds, except that they contain a few "bad" points where the shape appears folded or crinkled.

In a kind of cut-and-paste operation, Gordon and Webb showed how to transplant any possible vibration—imagined as a frozen wave—from one drumhead to the other. Because these waves can be reassembled successfully in either setting, the mathematicians proved that the characteristic frequencies of both membranes are the same. Although Webb was visiting Dartmouth University and Gordon was attending a meeting in Germany, the two managed to hammer out the final details of their proof via telephone and fax.

The mathematicians ended up with two mathematical drums of the same area and perimeter, each one a different multisided polygon, which resonate at exactly the same frequencies. In principle, two drums built out of these two different shapes would sound exactly alike. Thus, even someone who has perfect pitch wouldn't be able to distinguish the two sounds.

Since the initial discovery, Gordon and others have identified many pairs of soundalike drums, including sets that have an elegant simplic-

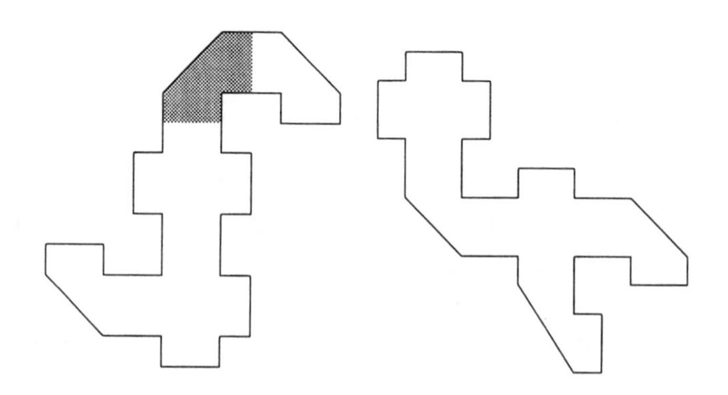

Although they are shaped differently, these two "drums" generate the same sound. In fact, each drum consists of seven half crosses (shaded unit) glued together in different arrangements.

ity. All known examples have at least eight corners, and, typically, each member of a pair consists of a set of identical "building blocks" arranged into different patterns. It's even possible to form different two-piece bands that make identical sounds. In this case, each band consists of two geometric objects, and this time the shapes are simple enough to allow a precise computation of the sounds they make.

Many issues remain unsettled, however, and new questions arise. For example, are there any soundalike triples—sets consisting of three geometric shapes, which taken together display identical normal modes? It's also known that not all drums have soundalike twins. For example, the only way to make a circular drum different is to change its diameter, and circles of different diameter make different sounds, so it isn't possible to have two different circular drums that sound alike. Whether there's a general way to tell which drums have twins and which don't isn't known.

In many ways, the detective work of deciphering what geometric information a spectrum holds has barely begun. So far mathematicians have been able to prove that most geometric surfaces are spectrally solitary—they have no isospectral twins. So, this soundalike quality is relatively rare. Most of the time, you do get distinguishable sound prints. Yet, in the orchestra of the mind, there is no escaping a delicious, disconcerting ambiguity in the beating of different drums.

Microwave Modes

When the physicist Srinivas Sridhar of Northeastern University heard about the Gordon-Webb-Wolpert discovery, he decided to put it to an experimental test. And he had just the right kind of setup to do the necessary experiment.

Sridhar and his coworkers had been investigating aspects of quantum chaos by looking at the patterns created when microwaves bounce around inside squat metal enclosures of various shapes. The same technique could be used to identify normal modes, with microwaves standing in for sound waves and severely squashed cavities standing in for membranes. In a way, it was like going back to Lorentz's original question about electromagnetic waves in a box.

To test the drum theorem, the researchers constructed two cavities corresponding to one of the pairs of shapes discovered by Gordon and her colleagues. Fabricated from copper, with eight flat sides, each

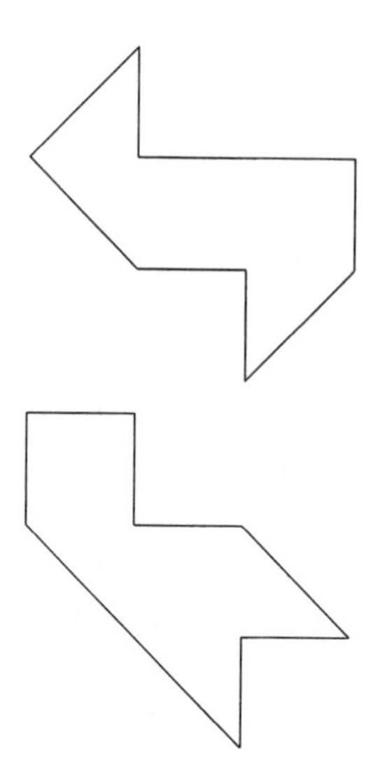

The simplest known pair
of soundalike drums.

angular enclosure was nearly eight centimeters long and less than six millimeters thick. Sending in microwaves through a tiny opening and measuring their strength over a range of frequencies at another location allowed the researchers to establish the normal modes of each cavity.

Remarkably, the experiment confirmed the mathematical prediction. The frequencies present in each spectrum were practically identical. Any discrepancies that were found could be attributed to slight imperfections introduced during the assembly of the enclosures. The experiment also provided information about the geometry of the normal modes themselves that was then unavailable mathematically (see color plate 2). "This interplay of mathematics and physics is beneficial to both fields," Sridhar and his colleagues concluded. "While the experiments have provided a satisfying physical basis for the mathematical results, the new ideas from mathematics which have been studied here may have wide and unforeseen impact on physical problems."

Later, Tobin A. Driscoll, using a supercomputer at the Cornell Theory Center, managed to compute the normal modes of the same pair of drums. His numerical results closely matched those of Sridhar and his colleagues.

The existence of isospectral shapes indicates that even in relatively simple settings, and with a complete set of measurements, it may be possible to reach more than one conclusion. This suggests that similar ambiguities could arise in various physical situations; for example, when geophysicists try to reconstruct Earth's interior from seismic data or when medical researchers generate images of internal organs from X-ray measurements. In other words, such systems may have characteristics that can't be determined from a measured spectrum.

Indeed, the same issue arises wherever the wave equation is applied, and its application covers a wide range of areas, from the formation and motion of water waves to the behavior of acoustic and electromagnetic disturbances. The equation's domain of applicability is so broad that it is often said to be the single most important mathematical equation ever devised.

Measurements of waves are routinely used in industry, for example, to test equipment for flaws. Any change in a component is often reflected in the way that it vibrates. So, by measuring its vibrations and looking for any changes, one can determine whether a component remains unblemished.

In the old days, an experienced railroad worker could walk down a rail line while striking the wheels of a stationary train with a metal bar and listening for suspicious tones. Workers knew the sound of a healthy wheel, and they could readily identify the abnormal sound of a wheel that was developing a crack. The same idea underlies a variety of modern, instrument-based methods for detecting hairline cracks and other defects, whether in the steel girders of a bridge or the titanium or graphite shaft of a golf club.

The drum results highlight several crucial mathematical issues involved in making judgments based on sounds, or, more generally, based on knowledge of normal modes of vibration. How many measurements do you need to get a reasonable answer? How much accuracy should you strive for?

Moreover, even if you could make infinitely many measurements with infinite precision, how sure could you be of your conclusions? Can two different objects appear alike in every measurable way? Interpreting the waves that pervade our lives requires answers to tough mathematical problems.

Fractal Tambourines

A glance at a map of the world reveals that coastlines are rarely smooth. More often than not, the rims of continents and large islands feature a wide range of projections and indentations. Gulfs, bays, fjords, peninsulas, spurs, and prominences, in turn, sport their own projections and indentations, and the pattern continues down to the irregular mazes of river deltas and the tumbled rocks of rugged shorelines.

Interestingly, the reason so much of a shoreline is ragged may have something to do with the interaction between water waves and edges. Waves rolling up and sweeping across a smooth beach dissipate their energy by churning up sand and eroding the shore. Waves broken up by a crazily indented coastline, which then crash into disordered heaps of rock, do much less immediate damage. Deeply indented coastlines may prevail simply because they can effectively dampen sea waves and therefore suffer less erosion.

The mathematics underlying this possibility involves the vibrational behavior of a membrane, this time not with a smooth rim but with an extremely crinkled boundary. Such irregular boundaries arise in many physical situations involving waves, from the jagged cracks in Earth's crust associated with earthquakes to the sound-modifying properties of textured walls in a modern concert hall.

One way to model such irregular boundaries mathematically is to use geometric forms called fractals. These are mathematical curves and shapes that look just as complicated when they're magnified as they do in their original form. A close look at such a figure reveals a miniature copy of the entire figure, and further magnification unveils even smaller copies. No amount of enlargement smooths away the structure.

Perhaps the archetypal fractal boundary is a curve first created in 1904 by the Swedish mathematician Helge von Koch (1870–1924). Like many shapes now known to be fractals, von Koch's curve can be generated by a step-by-step procedure that takes a simple initial figure and turns it into an increasingly crinkly form. This curve, which looks like a snowflake, starts as a large equilateral triangle. The addition of an equilateral triangle one-third the size of the original and pointing outward from the middle of each side of the large triangle turns the figure into a six-pointed star. The star's boundary has twelve segments, and the length of its outer edge is four-thirds that of the original triangle's perimeter.

In the next stage, a smaller triangle is added to the middle of each of the twelve sides of the star. Continuing the process by endlessly adding smaller and smaller triangles to every new side on ever finer scales produces the Koch snowflake. Any portion of the snowflake island magnified by a factor of three will look exactly like the original shape.

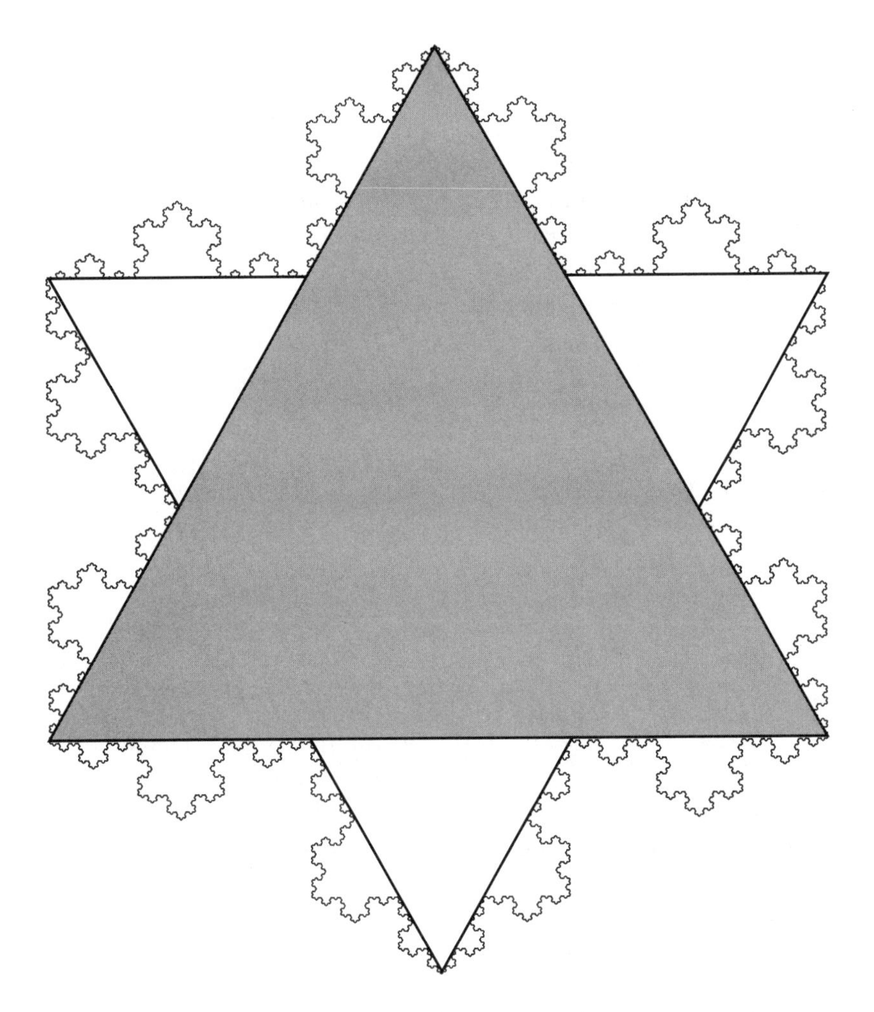

Starting with a large equilateral triangle, you can add smaller triangles to the middle of each side to create a six-pointed star. Adding even smaller triangles to the middle of each of the star's twelve sides generates a crinkly shape. Continuing the process by adding increasingly smaller triangles produces the intricately frilled Koch snowflake.

The boundary of the Koch snowflake is continuous but certainly not smooth. So convoluted that it's impossible to admire in all its fine detail, it has an infinite number of zigzags between any two points on the curve.

The inspiration for studying the vibrational frequencies of membranes with fractal boundaries comes from the work of Hermann

Weyl, who was particularly interested in the kinds of features all vibrating shapes have in common. He determined, for example, that the characteristic frequencies of oscillators typically fit a specific pattern defined by a simple mathematical relationship. The fact that the harmonics, or overtones, of a string are whole-number multiples of the fundamental frequency fits Weyl's law, as does the formula that gives the allowed frequencies of a circular drum.

Weyl's law applies to vibrating objects with smooth edges, and deviations from his law are largest for the lowest characteristic modes of vibration. In 1980, the mathematical physicist Michael Berry of the University of Bristol guessed that the amount of the error in applying Weyl's expression to complicated geometries depends on boundary effects. He proposed a refined version of Weyl's result that took into account how convoluted a particular boundary is. Thus, his proposed rule applied to shapes with smooth *and* irregular (or fractal) boundaries.

How would a drum shaped like a Koch snowflake vibrate? Berry supposed that it would vibrate much like a drum with a smooth rim, except in the fine details when the penetration of high-frequency vibrations into the tiny crevices of the boundary becomes important. The presence of ever-smaller nooks and crannies would allow more vibrational modes than a smooth-rimmed membrane, Berry conjectured. How many more and of what sort would depend on a quantity called the fractal dimension, which characterizes a boundary's crinkliness.

The most common way of defining a fractal dimension goes back to the work of the German mathematician Felix Hausdorff (1868–1942) at the beginning of the twentieth century. His idea was to consider dimension as a way of measuring the behavior of the Koch snowflake, or any other fractal curve, when its scale is changed. For example, suppose that, at the first stage of its construction, the snowflake curve had been one centimeter on each side. If the completed fractal were viewed with a resolution of one centimeter, the curve would be seen as a triangle made up of three line segments. Finer wrinkles would be invisible. If the resolution were improved to one-third centimeter, twelve segments, each one-third centimeter in length, would become evident. Every time the resolution was improved by a factor of three, the number of visible segments would increase four times. According to Hausdorff's formula, the Koch curve has a fractal dimension of about 1.2618, which is calculated by finding the natural logarithm of four and dividing it by the natural logarithm of three.

Berry's conjecture turned out to be false, but for an interesting reason: He chose an inappropriate (for this purpose) definition of fractal dimension. In 1988, the mathematicians Michel Lapidus and Jacqueline Fleckinger-Pellé proved a slightly modified version of the Weyl-Berry conjecture. To do this, they adopted a different, much more obscure way of defining the dimension of a fractal, which takes into account not only the convolutions of the curve itself but also the behavior of points near the curve. That's a reasonable choice for analyzing a vibrating membrane because it's the parts of the drum near its rim that are vibrating, rather than the rim itself.

In 1991, Bernard Sapoval and his coworkers at the École Polytechnique in Palaiseau, France, decided to look at the vibrations of a fractal membrane firsthand. The researchers cut a complicated pattern—resembling a cruciform snowflake—out of a stainless-steel sheet and stretched a plastic film across the opening as a membrane. A loudspeaker mounted above the "drum" excited the film, causing it to vibrate.

Sapoval and his coworkers observed that the membrane's convoluted boundary strongly damped excitations, soaking up their energy. Moreover, waveforms on such a surface showed extreme changes in slope at various locations along the membrane's rim. But these results applied to a complicated shape that is not really a fractal in the mathematical sense. A true fractal would have a never-ending sequence of tinier and tinier indentations—something that is not physically realizable. Did a surface with a truly fractal boundary show the same characteristics that Sapoval had found in his experiment?

Taking a mathematical approach, Lapidus, who is now at the University of California at Riverside, and his colleagues explored the characteristics of vibrations on membranes with fractal boundaries, concentrating on the Koch snowflake. The mathematicians proved several theorems, including one establishing that the edge of a membrane waveform indeed steepens sharply to an infinite slope as one gets closer to certain points along the rim. Many other conjectures, however, remained unresolved, and new questions kept coming up.

To gain insight into what may be going on and what kinds of mathematical questions are worth pursuing, Lapidus turned to computer graphics for clues. Working with a graduate student, Cheryl Griffith, and two collaborators, Robert Renka and John Neuberger of the University of North Texas, he brought displays of normal-mode vibrations of a drum with a fractal boundary to the computer screen.

Computer simulations of vibrations of a membrane shaped like a fractal snowflake reveal the first harmonic (top) and second harmonic (bottom). (Michel L. Lapidus, J. W. Neuberger, Robert J. Renka, and Cheryl A. Griffith)

The colorful images generated by Griffith vividly illustrate the dramatically frilled edges of the waveforms created on fractal-bounded membranes. Such fractal tambourines produce normal modes wrinkled in ways that reflect the influence of the membrane's boundary (see color plates 3 and 4). It also becomes evident that if one were to construct a fractal tambourine, it would have a distinctive sound. Its crinkled edge would muffle, even deaden, the tones, giving the tambourine's sound a subdued, muted flavor.

From a mathematical point of view, Griffith's numerical results couldn't convincingly depict all the nuances of a fractal membrane's vibrational behavior. As in a physical experiment, a computer can only approximate a fractal; it cannot render it in every detail. The resulting pictures must be interpreted carefully, and the underlying mathematics formulated and worked out. So, the studies continue.

Investigating the vibration of drums with fractal boundaries and drums with punctured membranes (flat surfaces perforated by an infinite array of holes) could, in the end, lead to a better understanding of such physical processes as the diffusion of oil through sand and the passage of waves through rubble.

One intriguing facet of fractal drums is what they may say about the apparent prevalence of irregular, fractal-like forms in nature, from crazily indented coastlines to the intricate branching of air passages in the human lung. For example, irregular (nearly fractal) coastlines may be more common than smooth ones because their shapes are fundamentally more stable. This would certainly account for the efficacy of disordered heaps of variously sized rocks as breakwaters. Other fractal structures may prove efficient in other settings (see chapter 8).

We venture now from the smoothness of individual waves and their interactions with edges, both even and irregular, to the complexity of their collective behavior and of the vibrating objects that generate them. We move from the harmony of pure tones to the randomness of noise.

How does the irregularity that characterizes noise arise from simple vibrations? One of the first experiments to show how pure tones generate noise occurred in 1952 at Göttingen, the city in Germany where Lorentz gave his lecture on electromagnetic waves in a box. When a liquid is irradiated with sound waves of high intensity, the liquid may rupture to form bubbles or cavities. This phenomenon is called acoustic cavitation and is typically accompanied by an intense emission of noise.

What the researchers at Göttingen found particularly interesting was how this emitted sound changes as the intensity of the pure tone that initiates the process slowly increases, starting from a low level. At first, the sound coming out of the liquid matches that of the input frequency. As the intensity increases, a second frequency abruptly becomes apparent, at half the original frequency. Then, as the intensity increases further, the frequency is halved again, so four different tones come out of the liquid. The halving continues until, at some threshold intensity, the cavitating liquid begins to generate what sounds like random noise. The sound generators in this case are oscillating bubbles, and their response to sound waves of increasing intensity gives rise to the phenomenon of acoustic chaos (see also chapter 7).

These observations take us into the realm of chaotic oscillations. In this domain, the features of a sound signal do not recur in a regular fashion; nor do they always change slowly. Moreover, the same thing can happen in all kinds of oscillators, from orbiting satellites, vibrating molecules, and light-generating lasers to beating heart cells and biological clocks.

The interplay of waves and noise—of order and randomness—is a winding, two-way road that ties together our lives, allowing us to begin to understand where we belong, yet ultimately limiting what we can know.

6

Noise Police

All nature is but art, unknown to thee;
All chance, direction, which thou canst not see;
All discord, harmony not understood . . .

—*Alexander Pope (1688–1744)*, "Essay on Man"

The compact disc's upper surface displays a stark, grainy portrait of a stern Ludwig van Beethoven (1770–1827). The shimmering, rainbow-hued surface on the opposite side holds a performance of his Ninth Symphony. In a CD player, the music comes to thundering life, translated from a vast array of microscopic pits, which cover the spinning disc's shiny, metallic face, into the sound-generating vibrations of stereo speakers.

Fresh out of its protective plastic case, the pristine disc delivers a brilliant, flawless sound. No pops, crackles, hums, skips, or jumps mar the glorious music. Months later, the disc bears the imprint of heavy use. Dust-catching, oily fingerprints track across its silvery surface. Even worse, the deep scratches of a cat's claws streak the gleaming finish. Yet, despite such damage, the CD continues to produce an unblemished sound.

A compact disc holds up to seventy-four minutes of music, just enough for a complete recording of Beethoven's Ninth Symphony on one disc. Every second of music is stored as a string of about 1.5 million bits, consisting of tiny, narrow pits, each ranging from 0.9 to 3.3 micrometers in length and representing the 1s and 0s of the digital signal that encodes the sound. It takes more than six billion bits, recorded on a spiral track more than 3.5 miles long but only 0.5 micron wide, to capture the entire symphony.

When the CD player's laser beam, focused on the surface of the revolving disc, strikes one of the microscopic molded pits, its light is scattered. The level of light reflected back to a detector changes. Nudged by the detector, the instrument's electronic circuitry interprets the transitions from depression to smooth surface as binary 1s. Intervals of different lengths between these abrupt transitions are spelled out as the corresponding number of 0s. Strung together, these bits carry the information needed to re-create the sounds of an orchestra.

Such a process, however, is quite susceptible to error. A speck of dirt, a fingerprint, a scratch, or some other defect can readily hide or alter the disc's minuscule surface features. To circumvent this problem, recording engineers add extra bits to the embedded signals, taking advantage of mathematical procedures that not only detect errors but also repair them. In effect, these error-correcting schemes distinguish between random noise and information-carrying bits, reinforcing patterns and restoring meaningful signals that could otherwise get lost or blurred.

Over the last fifty years, mathematicians have invented a variety of ingenious strategies to ensure that a CD player, a computer, or any other device that receives or handles digital information can automatically detect and correct random errors. Mathematically based error-correcting codes now permit clean, crisp sound reproduction even from a scratched or dirty compact disc. They also help to assure accurate storage of data on a computer's hard drive and reliable data communication at low power over long distances via cellular phone, fax machine, modem, or satellite.

Error-correcting codes represent a mathematical triumph over randomness. But their magic is limited. A determined child, with screwdriver or paper clip in hand, can wreak sufficient havoc to damage a CD beyond recovery.

Word Play

Error detection and correction is something that we do all the time, often so effortlessly and naturally that we don't realize we're doing it. Suppose you're reading the want ads in a newspaper and you come across the phrase "must have a minimum of bive years' experience." The word *bive* stops you for an instant. You're not familiar with it, and you suspect a printing error. Fortunately, the rest of the phrase contains enough information for you to guess that the word ought to be *five*, and you continue reading.

Of course, that's not the only possibility. With just one incorrect letter, the word could have been *dive* or *bide*. If two letters were wrong, there's a chance that it could have been *fire* or *bird*. The context, sentence structure, and natural redundancy of the English language, however, usually limit the options to a reasonable few.

Consider the following five-letter words, each of which contains an incorrect character: *qsiet*, *tradf*, *thara*, *splir*, and *scout*. Because there is a pattern, however subtle and complex, to the way words are constructed in English, it's possible to make good guesses about what some of these words are meant to be. For example, it's exceedingly rare for any letter other than *u* to follow *q*, so the first word must be *quiet*. Many words end with this pattern: vowel, consonant, silent *e*. The second word should be *trade*. The patterns are trickier to pick out in the next two examples, but in each case only one possibility makes sense: *tiara* and *split*.

The final example, however, presents all sorts of problems. It looks correct to start with. But if one letter really is incorrect, then the word could be *scour.* Because it's not clear which letter is wrong, it could also be *shout, snout, spout, stout,* or *scoot.* Spell-checking software faces such dilemmas all the time, and the only reasonable action it can take is to present the writer with a list of possible replacements. That works reasonably well—unless an error happens to masquerade as a legitimate word.

"English has quite a lot of built-in error-correcting ability because of its natural patterns," says Robert McEliece, an electrical engineer at the California Institute of Technology, who is well known for his work in applying mathematics to various problems in communications theory. "But it wasn't designed systematically to correct errors, and . . . sometimes changing one letter in a word can dramatically change its meaning. Of course, it is just this wonderful ambiguity that lets us play word games, commit horrible puns, write complex poetry, and invent pseudo-words like 'mathemagically.' "

Such ambiguity, however, is not acceptable in the supermarket, with an airline, or at a bank. An incorrect word, a misplaced digit, or a missing letter can spell trouble. At the very least, businesses, financial institutions, government agencies, and other concerns need a way of quickly flagging a large proportion of such mistakes, even if correcting them automatically is simply not feasible.

Cereal Boxes and Airline Tickets

You lift up the telephone receiver, punch in the numbers 9 and 1 and the ten digits of a long-distance number, wait for the chime, then punch in the fourteen digits of the calling-card number. A moment later, you hear: "The card number you have dialed is not valid." You look closely at your telephone's digital display and you find that indeed you have entered one of the calling-card digits incorrectly. You try again, and the call goes through.

How did the telephone company flag the incorrect card number, thus ensuring that the call would not be billed to the wrong person or company?

The secret lies in the use of fast, reliable computers to perform quick calculations on the entered number, following mathematical recipes that pinpoint mistakes. Such methods apply to a wide range of identification numbers, from the bar codes used to tag groceries or rental cars to the lengthy string of numbers on credit cards, airline tickets, and driver's licenses. The assigned string of digits not only serves as an identifier but also typically incorporates an extra "check" digit as a means of detecting forgery or error so that one product or person isn't mistaken for another.

Various methods are used to generate check digits in commercial situations. For example, the last digit of an airline ticket number should equal the remainder left after the rest of the digits are divided by 7. Thus, the ticket number 17000459570 3 passes this test because dividing the number by 7 leaves 3 as the remainder: $17{,}000{,}459{,}570 = 7 \times 2{,}428{,}637{,}081 + 3$.

Applied to a garbled version of the original number, this simple recipe catches many, though not all, possible errors. For instance, the method readily detects most transpositions of adjacent digits. Thus, if the number 5400474 is accidentally entered as 4500474, the computer would calculate that the check digit should be 3, which doesn't match the 4 at the end of the number—so there's an error someplace. However, it doesn't work for transpositions of the pairs 0 and 7, 1 and 8, and 2 and 9. The method also doesn't catch the substitution of 0 for 7, 1 for 8, 2 for 9, and vice versa at any position along the string.

A more complicated strategy, commonly applied to bar-coded numbers used to identify products, detects a larger proportion of errors. For example, a box of Kellogg's Corn Flakes has the following number 0 38000 00127 7. According to the Universal Product Code (UPC), the first six digits identify the manufacturer, the next five tag the product, and the last one is the check digit.

Suppose the laser scanner at the local supermarket reads this bar-coded number as 0 58000 00127 7. How does the computer connected to the scanner detect the error?

The computer is programmed to add together the digits in positions 1, 3, 5, 7, 9, 11 and triple the result, then add this tally to the sum of the remaining digits. If the result doesn't end with a zero, the computer knows that the entered number is incorrect. In this case, $[(0 + 8 + 0 + 0 + 1 + 7) \times 3] + (5 + 0 + 0 + 0 + 2 + 7) = (16 \times 3) + 14 = 62$, and the result doesn't end in 0, so the error is detected and the cashier hears a beep. In general, this particular scheme detects all substitution errors at any single position and nearly 90 percent of other possible mistakes.

Many credit-card issuers have adopted a slightly different method, which catches 98 percent of the most common errors. For example, American Express traveler's checks use a check digit chosen so that the

Anatomy of a bar code: The first digit of the main block identifies the kind of product. For example, a 9 indicates that the item is a book. The next six digits represent the manufacturer. The following five digits are assigned by the manufacturer to designate the product, including size, color, and other important information (but not the price). The final number is the check digit, which signals the computer if one of the other digits is incorrect.

sum of the digits, including the check digit, is evenly divisible by 9. Thus, a traveler's check with the serial number 210687315 must include 3 as its final check digit because 2 + 1 + 0 + 6 + 8 + 7 + 3 + 1 + 5 = 33, and 33 + 3 is evenly divisible by 9.

The method of assigning check digits that a company chooses represents a balance between the need for performing checks as quickly as possible and the types of errors that it would like to catch. In some cases, the schemes must handle both numbers and letters of the alphabet. In other cases, it's enough to perform just a parity check, which involves determining whether the number in question is odd or even. Parity checks are particularly convenient when information is in the form of binary digits.

Error correction goes one step further by providing a means of not only identifying a mistake but also replacing it with the correct digit or symbol. Over the years, researchers have developed a number of schemes for performing this crucial cleanup operation.

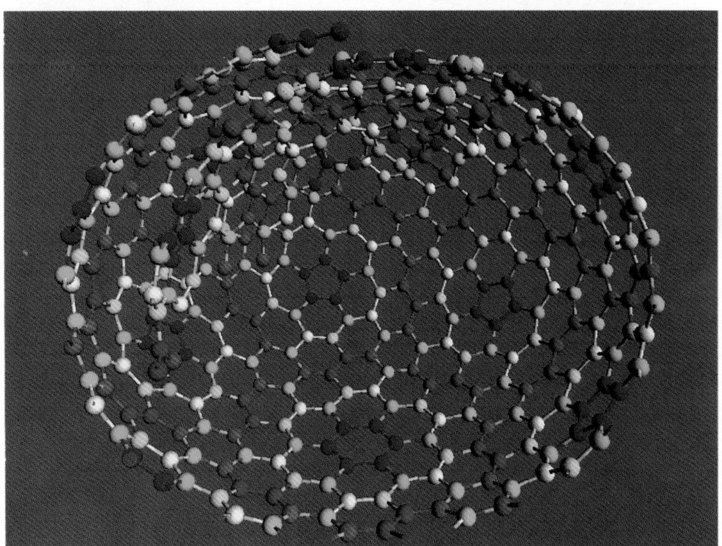

Color Plate 1 **Shell Rules:** A computer simulation demonstrates that an icosahedral virus shell (top) can form spontaneously if its protein building blocks fit together only in certain ways specified by a set of seven lock-and-key rules (see page 58). A single misstep in the building process can lead to a spiraling defect, which keeps the virus shell from closing (bottom). (Courtesy of Bonnie Berger, MIT)

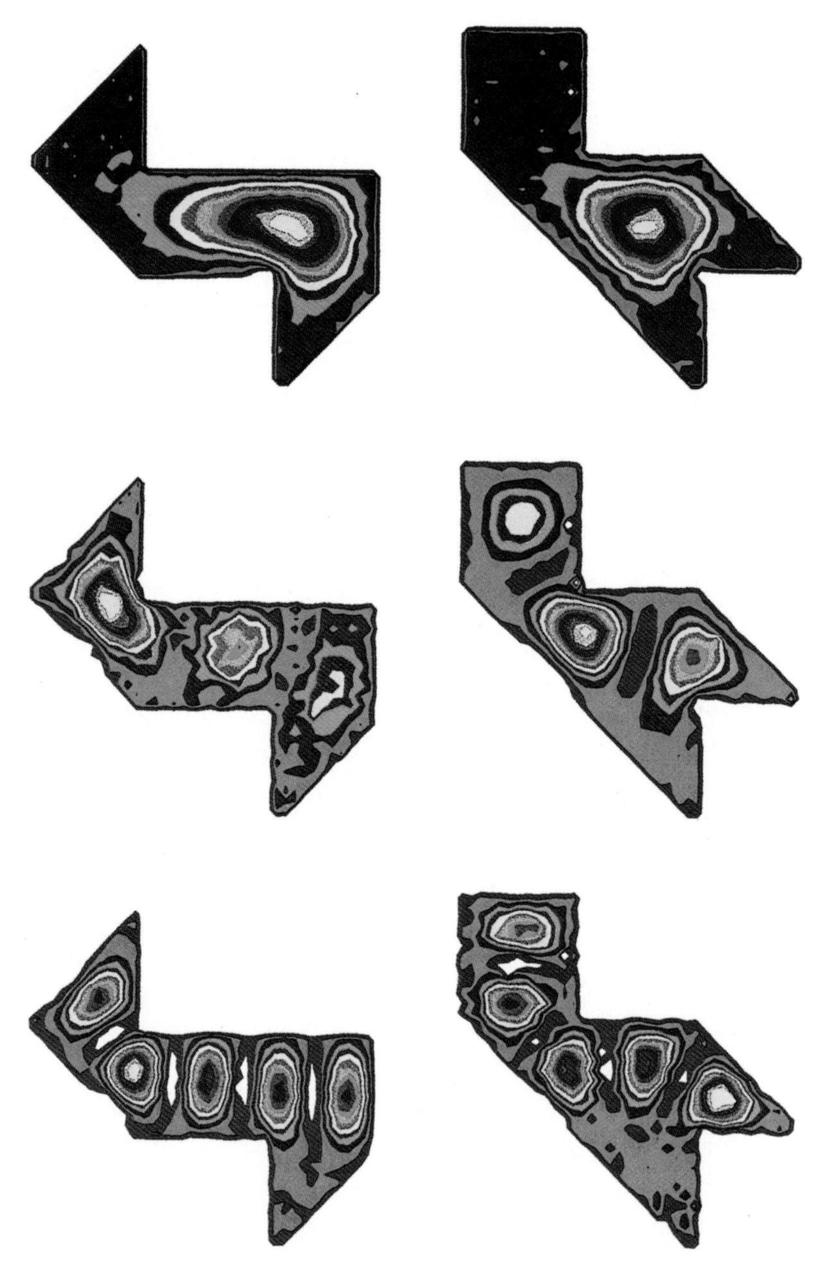

Color Plate 2 **Drum Harmony**: Pumping microwaves into a thin metal box with a top surface shaped like each of two different drum geometries reveals that both have normal modes of the same frequency, providing experimental confirmation that the two drums sound alike (see page 100). These plots reveal the geometry of the standing waves of the first few normal modes. (S. Sridhar, Northeastern University)

Color Plate 3 **Snowscape:** Mathematical studies reveal intriguing details of how a drum's shape affects its sound, even when its shape is a fractal snowflake curve (see page 106). Using computer graphics, mathematicians can visualize vibrations of a fractal drum's membrane. This computer-generated image displays one of its normal modes of vibration. (Michel L. Lapidus, J. W. Neuberger, Robert J. Renka, and Cheryl A. Griffith)

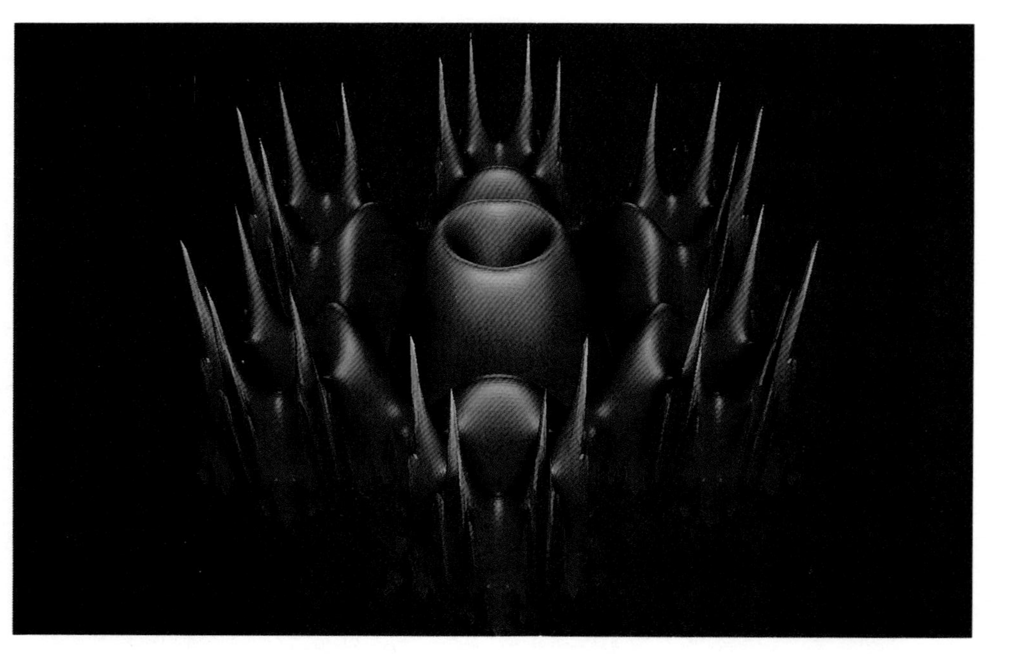

Color Plate 4 **Spiderflake:** When a membrane shaped like a fractal snowflake vibrates, it curves into a set of hills and valleys (see page 106). The slopes of such surface features can vary greatly, as shown in this computer-generated depiction of the membrane steepness at different points on the surface at a given moment during one vibration. (Michel L. Lapidus, J. W. Neuberger, Robert J. Renka, and Cheryl A. Griffith)

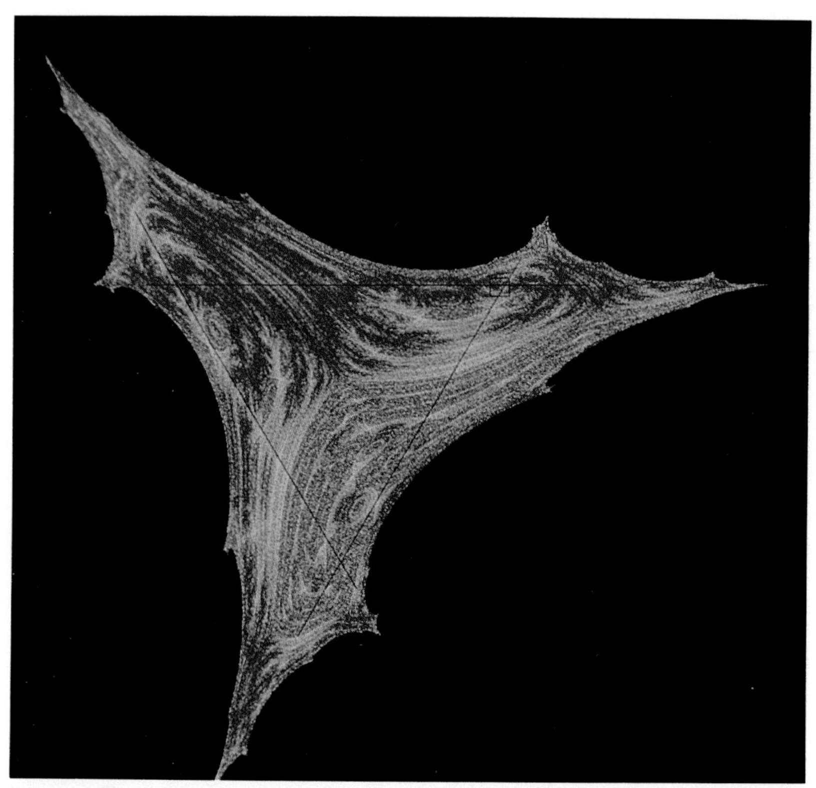

Color Plate 5 **Fractal Foam:** The intricate lacework of colors in this computer-generated portrait of a dynamical system illustrates an unusually wild type of behavior that sometimes arises from the manipulation of simple mathematical expressions (see page 139). The particular set of equations considered here has three attractors (three lines that intersect to form a triangle); the color (red, green, or blue) signifies to which one of the three attractors a given starting point will go. Magnifying a portion of an intermingled, riddled basin reveals that neighboring points readily end up on different attractors. (James C. Alexander, University of Maryland)

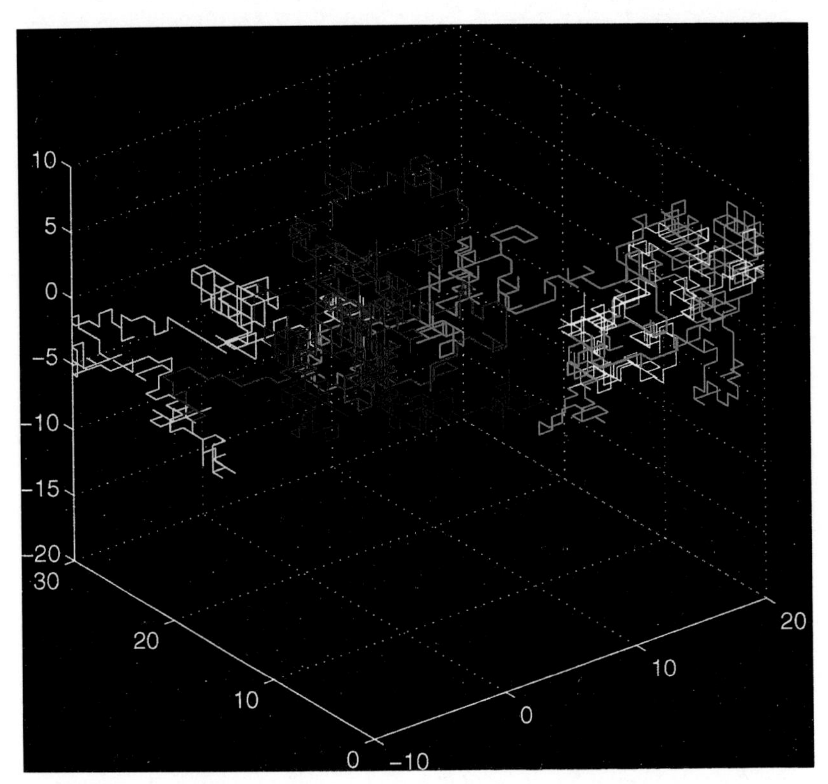

Color Plate 6 **Lost in Space:** In this computer-generated representation of a random walk in three-dimensional space (see page 155), a walker has an equal probability of moving forward, backward, right, left, up, or down. This particular walk goes for 2,100 steps, beginning with blue, then continuing with magenta, red, yellow, green, cyan, and white segments. Extended indefinitely, the walk has only a 34 percent chance of ever returning to its starting point. (G. M. Viswanathan)

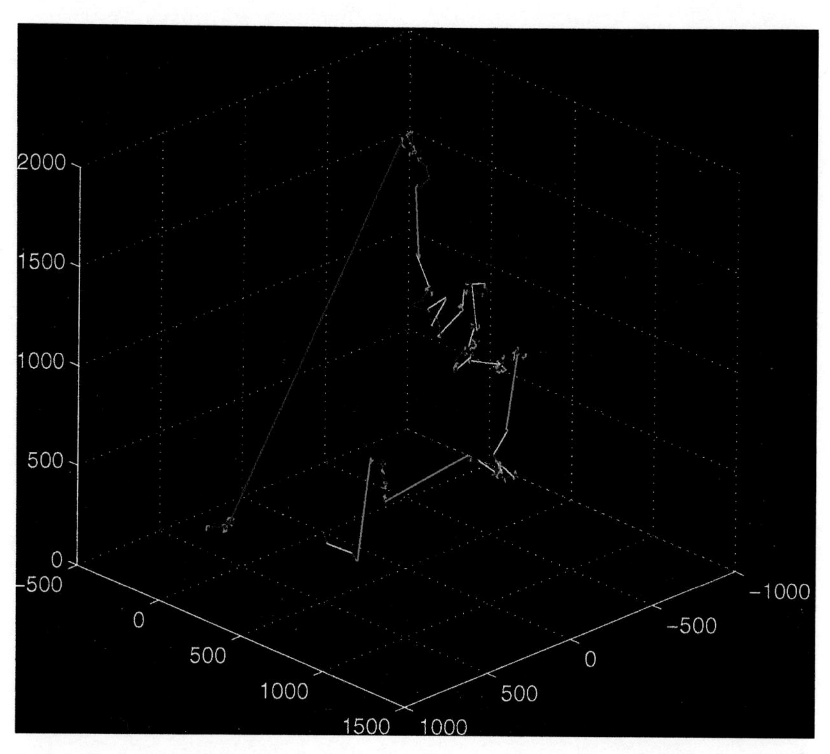

Color Plate 7 **Flight Time:** A Lévy walker takes steps of different lengths, with longer steps occurring proportionally less often than shorter steps (see page 162). In three dimensions, Lévy flights correspond roughly to a sequence of long jumps separated by what looks like periods of shorter ventures in different directions. Each of these periods of shorter excursions, in turn, is made up of extended flights separated by clusters of shorter flights, and so on. Magnifying any of the clusters or subclusters reveals a pattern that closely resembles the original large-scale pattern, which means that Lévy flights have a fractal geometry. The illustrated flight goes for two thousand segments. The color of a step depends on its length. (G. M. Viswanathan)

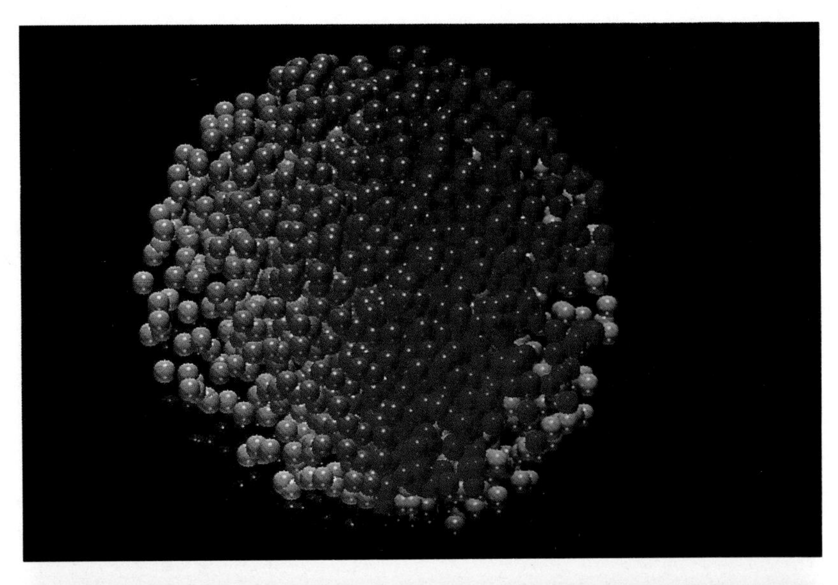

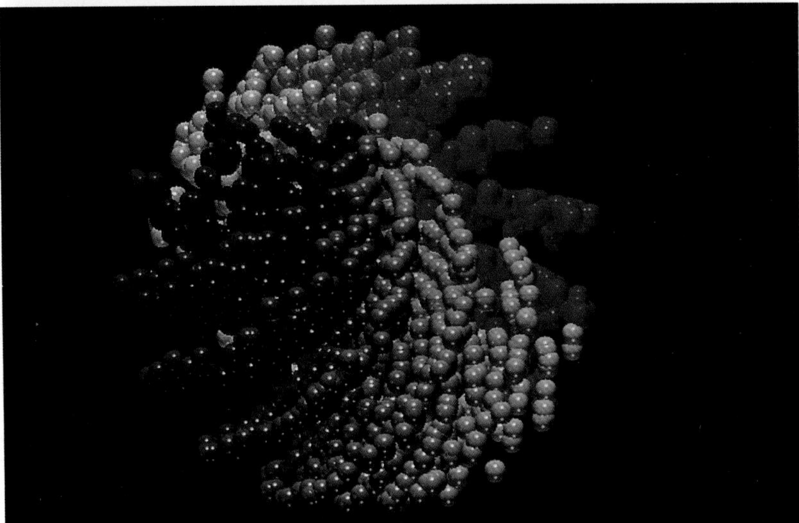

Color Plate 8 **Noise Sphere:** One way to check the randomness of the output from a random-number generator is to present the data in visual form, making any patterns that may exist easier to pinpoint (see page 180). In this case, three consecutive "random" numbers are used to specify the position and color of a small ball in three-dimensional space. Successive triples produce a spherical distribution of colored balls. Collections of balls representing the output from high-quality random-number generators show no discernible patterns (top), whereas low-quality generators yield tendrils and other distinctive features (bottom). The technique is remarkably sensitive to small deviations from randomness. (Clifford A. Pickover, IBM Research)

Flipped Bits

Communication is generally imperfect. Even if a message is accurately stated, a noisy environment—whether it is static disturbing a radio broadcast or voltage spikes punctuating an electric signal moving down a wire—can interfere with its transmission and potentially alter its meaning. To get communication of the highest possible quality, it helps to design and use a precisely defined language. One simplification is to step away from the twenty-six-letter alphabet of the English language to the realm of 0s and 1s. Limiting the basic elements of the language, or code, to just two has a distinct advantage in that it's generally easy to represent and distinguish the two states.

A century ago, Morse code was a quick, reasonably efficient way of sending information over long distances. A telegraph operator tapped a key that opened and closed a switch to generate a string of electric pulses encoding the news, which was then transmitted over wires to its destination. The dots and dashes of the code became short and long pulses of electric current, which had to be separated by brief intervals of no current to ensure intelligibility.

Nowadays digital communication involves similar signals, in the form of bits, to convey information. Here, the signals are typically sent in blocks, consisting of a given number of bits each, to preserve clarity. However, static and other types of environmental noise can sometimes flip a bit, so 1 comes through as 0, and 0 as 1. Suppose, for instance, that the sequence 00101011 means *plus*. It takes only two flipped bits to alter the original sequence to 00101101, which might stand for *minus*. Addition suddenly becomes subtraction!

Such errors can have disastrous consequences when financial records, diplomatic missives, military signals, and other sensitive messages end up being mangled and misinterpreted. The role of error detection and correction, then, is the maintenance of pattern in the face of noise.

A person speaking over a crackly telephone line has several options for making himself understood at the other end—for example, he can talk louder or more slowly. And if the noise persists or proves excessive, he can repeat his words several times or even spell them out. People at large parties or raucous gatherings often adopt similar strategies.

All forms of communication require a certain amount of energy to produce the electric current variations, radio signals, light flashes, or sound waves that transmit an intelligible message. However, just as it isn't always possible to yell louder at a party to get a message across, it doesn't always make sense to boost power to transmit information successfully.

The key to digital error detection and correction lies in redundancy. For example, one can simply repeat each of the digits of a message a specified number of times. Thus, the message 101 can be transmitted as 111000111, and the computer at the receiving end would decode this sequence to get 101. If one bit changed during transmission, two others would still be correct, and the computer would select the majority digit as the appropriate entry. The trouble is that this method requires the transmission of a large number of extra digits, making it slow and inefficient.

In the late 1940s, in a landmark paper on the mathematics of communication, Claude E. Shannon, who was then an applied mathematician at Bell Telephone Laboratories, showed theoretically that for digital messages, there are methods of coding the data that do much better than simple repetition to ensure accuracy when the message is transmitted over a noisy channel. However, his theorem indicated only that good codes exist. It did not specify what they would be like or how complicated they would be to use.

One approach to developing such an error-correcting code requires adding a small number of check digits to the end of each segment, or block, of a message. The procedure resembles the verbal ploy of saying "a" as in "alpha," "r" as in "Romeo," and so on, to help a listener correctly identify the letters of a spelled-out word. The trick is to add just enough bits to guarantee a desired level of reliability without adding so many bits that it takes a prohibitively long time to get the message through. A CD player that processes extra digits to clean up a disc still has to play the music without introducing a noticeable stutter.

Hence, users of error correction must balance maximizing the reliability of data transmission against keeping the rate of information transfer as high as possible. They also need to consider the overhead imposed by the time-consuming encoding and decoding operations required before and after data transmission. Mathematical ingenuity comes into play in this trade-off between accuracy and efficiency.

Glitch Patrol

Early computers were notoriously unreliable. One of the pioneering machines of the 1940s was an electromechanical calculator at Bell

Telephone Laboratories. Occupying about a thousand square feet of floor space and weighing more than ten tons, this hulking giant relied on more than nine thousand metal switches, which could be forced back and forth between two positions by means of electrically controlled magnets, to perform calculations.

The machine was very good at detecting errors. It simply stopped when it suffered a mechanical or electric fault or, if the error was in a program of instructions, switched to the next program waiting in line. Such contrary behavior, however, proved immensely frustrating at times, and it goaded Richard W. Hamming (1915–1998), a mathematician at Bell Labs, into developing a way for the computer to correct certain types of program errors and proceed without human intervention.

The year was 1947, and Hamming had access to the Bell Labs computer only on weekends. Instructions were entered into the machine by means of a long strip of punched paper tape, which was nearly an inch wide. Up to six holes could be punched in each row spanning the tape's width, and each of these rows was read as a unit. Different patterns of holes represented different instructions.

The system was designed to check itself during computations. For example, if the sensing relays detected more or fewer than two holes in a given row, the machine would abruptly stop the computation and sound an alarm. The operator could then locate the problem by the pattern of lights on an elaborate panel, which signaled the machine's status at the moment computation had stopped. At night or on weekends, however, there was no one on duty, and the machine was rigged to dump a program if it encountered an error. "Two weekends in a row I came in and found that all my stuff had been dumped and nothing was done," Hamming later recalled. "And so I said, 'Damn it, if the machine can detect an error, why can't it locate the position of the error and correct it?' "

Out of that frustrating setback came a novel, elegant method of coding information to allow for remarkably efficient error correction. Instructions or messages are typically coded in blocks, represented as strings of 0s and 1s of a given length. For example, the twenty-six letters of the English alphabet can be coded using blocks that consist of five binary digits, starting with 00001 for A, 00010 for B, 00011 for C, and so on. A sequence of such blocks constitutes a message: 01110 00101 10111 10011. In general, the different blocks belonging to a given set are called codewords, and the list of all allowable words in such a set is known as a code.

When errors never occur during data transmission or storage, all such blocks may be used to form messages without fear of misinterpretation. If errors do occur, however, the blocks received may differ from those that are sent. One way to compensate is to send only the types of blocks that differ from each other so much that one or two errors can't change one of the selected blocks into another legitimate codeword.

Consider the simplest possible code, which has only two codewords: 0 and 1 (meaning, for instance, yes and no). No error correction is possible with this code because any mistake changes one codeword into the other. Using the codewords 00 and 11, formed by repeating the digits of the original codewords, allows the possibility of detecting single errors, because such a mistake changes a given codeword into one that doesn't belong to the list of allowed strings. Correction, however, isn't possible.

Suppose the two codewords are 00000 and 11111, and a message that starts out as 00000 is received as 01000. If the transmission channel isn't too noisy and such errors aren't particularly likely to occur, the message recipient is certainly justified in deciding that the original codeword must have been 00000. In a sense, the decoder chooses the codeword that is closest to the received sequence. The noisier the transmission channel, the longer a string of 1s and 0s the parties must use for each codeword to assure accuracy.

Hamming exploited the idea of the "distance" between codewords to develop an effective error-correction scheme. He defined such a distance as the number of places in which two codewords differ. For example, the distance between 01000 and 0000 is one because the two strings differ in only one place. In contrast, the distance between 01000 and 11111 is four. For efficient error correction, the trick is to find a set of codewords where the distances between them are as large as possible, balanced against the need to minimize codeword length to reduce transmission time.

One code that Hamming designed requires seven digits in each codeword. The message itself is conveyed in blocks of four digits each, ranging from 0000 to 1111. Three carefully selected digits are then added to each starting block to provide the maximum possible distance between codewords. The resulting set has such a strong pattern that random flips of one or two bits in any block would stand out like a sore digit.

Thus, in the Hamming code 0000 becomes 0000000, 0001 becomes 0001110, 0010 becomes 0010101, and so on, right up to 1111, which becomes 1111111.

Hamming Code

Message	*Codeword*	*Message*	*Codeword*
0000	0000 000	1000	1000 111
0001	0001 110	1001	1001 001
0010	0010 101	1010	1010 010
0011	0011 011	1011	1011 100
0100	0100 011	1100	1100 100
0101	0101 101	1101	1101 010
0110	0110 110	1110	1110 001
0111	0111 000	1111	1111 111

This example of a Hamming code requires four binary digits to represent each different letter, symbol, or message and three extra bits to allow for error correction.

Consider the message 1000 0101. Encoded, it reads: 1000111 0101101. After a transmission error, the message could look like this: 1000011 0001101. The first block doesn't exist among the codewords in the Hamming scheme. But it differs from the codeword 1000111 in only one place. No other codeword comes this close, so the decoding computer selects this codeword as the most likely possibility and corrects the flawed block appropriately.

Of course, the computer may still make the wrong selection, depending on the severity of the digit scrambling that occurs during transmission. But by choosing the extra digits in each codeword carefully, communications experts can significantly reduce the probability of such cases of mistaken identity.

Discs and Drives

Designing a code is like building tidy pyramids of oranges in the grocery store. Both activities involve piling things up into the tightest or most efficient arrangements possible without squeezing them out of shape.

Codewords must be as short as possible to minimize transmission time and conserve power, yet long enough to ensure a minimum distance between permissible choices. One can picture this minimum distance as the diameter of a rigid sphere, and the number of additional spheres that can be packed around a given center determines the number of possible codewords.

One can perform this sphere-packing in higher, impossible-to-visualize dimensions. Suppose, for example, that you need sixty-four codewords (to represent, say, sixty-four different colors in an image). Without redundancy for error correction, it would take six bits to represent all the colors.

To make sure there is enough room between the codewords, one can envision the original six-bit codewords as hypothetical, six-dimensional spheres, each one having an address, located in a six-dimensional box, given by its digits. One can then imagine increasing the number of dimensions—adding extra digits to each address—until enough numerical space opens up between the imaginary, higher-dimensional spheres to allow an acceptable level of error correction.

Suppose that happens when the codewords are ten digits long. The addresses locating the centers of ten-dimensional spheres in the tightest possible packing give the most efficient codewords.

A similar strategy for defining suitable sets of codewords involves finding optimal arrangements of points on the surface of a sphere. At first glance, the problem seems trivial. But upon closer inspection its intricacies become evident. For instance, suppose fifty fiercely competitive owners of pizza parlors on a newly colonized, oceanless planet want to situate their establishments as far apart as possible. This is an example of the packing problem: placing a given number of points on the surface of a three-dimensional sphere so that the points are separated by the largest possible distance. Eager customers, however, look at the situation a little differently. They want the parlors positioned so that no matter where someone lives, the distance to the nearest establishment is minimized. This is an example of the so-called covering problem.

These scenarios represent just two of at least ten different criteria for choosing the position of points on a sphere's surface. For a given number of points, the resulting geometric arrangements generally differ from case to case, and the trick is to choose the right criterion for a given application, whether it is error correction or some other function. Moreover, the problem of distributing points on a surface can be extended to the analogs of spheres in four and higher dimensions.

Neil J. A. Sloane and Ronald H. Hardin of Bell Labs, along with Warren D. Smith, have spent years building up extensive tables of arrangements of points on spheres. Their work has contributed to solving a wide variety of problems in areas ranging from digital communication to experiment design and numerical analysis.

Mathematicians and others are constantly on the lookout for practical error-correcting codes that are both more efficient and more compact than those currently in use. One particularly important family of such codes is the Reed-Solomon code, which is used for storing information on a computer's hard drive, processing digital images, and ensuring a faithful reproduction of the sound on a compact disc.

Developed in 1960 by Irving S. Reed and Gustave Solomon, then members of MIT's Lincoln Laboratory, one example of this code can correct up to four thousand consecutive bit errors. That means you can drill a hole three millimeters wide in a compact disc and still suffer no loss in sound quality. The basic idea underlying the scheme is geometric. The digital data are encoded as part of a pattern so strong that it can be recognized even if part of it is missing. It's like recognizing the original shape of a chunk of cheese known to be precisely cubic after mice have nibbled away parts of it.

Suppose that the data consist of two numbers: 3.6 and 5.9. Encoding them according to the Reed-Solomon method is roughly equivalent to plotting them on a sheet of graph paper, using the points (1, 3.6) and (2, 5.9). Those points define a straight line. One can then pick four additional, equally spaced points that fall along the same line, such as the points (3, 8.2), (4, 10.5), (5, 12.8), and (6, 15.1), leading to the following sequence of numbers: 3.6, 5.9, 8.2, 10.5, 12.8, and 15.1.

Suppose that two errors occur to change the set to the following numbers: 3.6, 5.6, 8.2, 10.6, 12.8, and 15.1. When those numbers are plotted, the straight-line pattern is spoiled but not obliterated. Four of the six points still fall on the original line, which can then be picked out. To recover the data, all we need to do is move the wayward points back into line. Of course, this particular example won't work if there are three errors. In that case, one would need six additional numbers to provide sufficient redundancy to do accurate error correction.

In practice, one sends not just pairs of numbers but three or more numbers at once. Instead of straight lines, more complicated curves are needed. For instance, three numbers can be used to define a bowl-shaped curve called a parabola. Picking four additional, evenly spaced points along a parabola defined by the three given data points again produces a strong pattern. It's extremely unlikely that seven numbers chosen at random would all lie on the same parabolic curve.

The Reed-Solomon method is particularly good at dealing with bursts of errors, like those caused by a particle of dust or a scratch on the surface of a compact disc. On a CD, the six billion bits encoding the sound are accompanied by about two billion additional, error-correcting bits. The sophisticated Reed-Solomon decoder of a typical CD player processes about two million coded bits per second. Thanks to error correction, the music you hear on a CD isn't merely almost as good as the original; it's precisely as good as the original.

Cell-Phone Codes

The sight of someone talking into a cell phone while driving a car or walking down the street is no longer a rare occurrence. The United States alone has more than eight million cellular-telephone users, and demand for cellular service has grown so rapidly that providers have been forced to look for new ways of increasing the capacity of their overtaxed systems.

A cellular telephone is basically a mobile two-way radio. In a cellular system, an area is divided into smaller sections called cells. Each cell contains its own antenna, which handles signals traveling to and from any mobile phone within its reach. Because the signals are relatively weak, the same radio frequencies can be used in different cells spread out over a given area.

However, each cell has a limited number of radio channels available for carrying telephone conversations. To increase system capacity, especially in cities, network operators have made cells smaller while studying alternative methods of transmitting signals. One such technology is known as code division multiple access (CDMA), which allows many users to broadcast over the same channel simultaneously.

In the CDMA approach, the system keeps signals straight by assigning a separate codeword or sequence of digits—like those used in error-correcting codes—to each user as an identifying tag. Increasing the number of codewords permits more users to obtain access to the system. It was while studying ways to generate longer sequences of CDMA codewords that P. Vijay Kumar of the Communication Sciences Institute at the University of Southern California uncovered a remarkable, previously unsuspected mathematical link between two types of error-correcting codes that were originally thought to be quite distinct.

To be practical, error-correcting codes must be designed so that simple computer routines can easily generate and recognize encoded information. So-called linear codes, which exhibit strong patterns, meet this criterion.

In general, adding together any two codewords of a linear code generates another codeword from the list. Consider, for example, the simple linear code that consists of the following codewords: {00000, 11100, 00111, 11011}. Using the binary addition rules 0 + 0 = 0, 0 + 1 = 1 + 0 = 1, and 1 + 1 = 0, one can easily show that 11100 + 11011 = 00111. Linearity gives a code a mathematical structure that makes encoding and decoding messages a snap.

Nearly all error-correcting codes now in commercial use, including the Hamming and the Reed-Solomon codes, are linear. However, researchers have long known that nonlinear codes of a given length can have more codewords than their linear counterparts. But decoding nonlinear codes has proved to be so cumbersome and complicated that using them hasn't been worthwhile.

Kumar and his collaborators started by creating a potentially useful, efficient linear code based not on binary digits (0 and 1) but on quaternary digits (0, 1, 2, and 3). Expressed in quaternary digits, 0 = 0, 1 = 1, 2 = 2, 3 = 3, 4 = 10, 5 = 11, 6 = 12, 7 = 13, 8 = 20, and so on. Quaternary arithmetic follows the rules: $2 + 2 = 1 + 3 = 2 \times 2 = 0$ and $3 + 3 = 2 \times 3 = 2$.

The surprise came when the researchers expressed their new linear quaternary code in binary form. That is, they replaced each 0 of the quaternary code with 00, 1 with 01, 2 with 11, and 3 with 10 to get binary strings that were twice the length of the original quaternary strings.

Kumar and Roger Hammons, a USC graduate student at the time, recognized in the binary strings the so-called Kerdock code, one of the well-known, highly efficient nonlinear codes that had previously proved so difficult to use. Similar relationships linked quaternary sequences with other nonlinear codes. Expressed in quaternary form, the nonlinear codes became linear and relatively easy to use.

Meanwhile, unaware of the work that Kumar and Hammons had done, Neil Sloane, Rob Calderbank, and Patrick Solé, working at Bell Labs, started a little later and followed a different mathematical track to come to the same conclusion. Calderbank found out about the USC results when he became aware of the title of a talk by Kumar and Hammons scheduled for presentation at an information-theory symposium. He telephoned Kumar, and the researchers discovered a great deal of overlap in their results, although each group had chosen to elucidate different details. They decided that they would be better served by writing one paper, rather than two, describing the discovery.

Several groups are now going beyond the quaternary digits to other sets of numbers in hope of finding even better codes. They are also investigating decoding techniques that take advantage of the newly discovered relationship between binary and quaternary codes. As the understanding of the mathematical underpinnings of linear and nonlinear codes steadily grows, we can look forward to exchanging messages using pocket communicators that take advantage of the new codes to keep airborne messages straight.

Secret Missives

Modern technology has made eavesdropping—whether officially sanctioned, inadvertent, or illegal—remarkably easy. Today, cellular and cordless telephones transmit conversations via radio waves that can be readily intercepted. Electronic-mail messages pass openly from one computer to another across a network that is accessible to innumerable computer users.

However, with a little encryption to hide their words, Prince Charles, Princess Diana, and other unwitting celebrities and politicians might never have suffered the embarrassing spectacle of having transcripts of their private telephone conversations splashed across the front pages of newspapers around the world. Whereas error correction cleans up a message, encryption scrambles it, turning digitized sounds and text into gibberish. Only a recipient who is equipped with the proper key for unlocking the secret code can hear or read the encrypted string of digits.

One way to achieve privacy is for both the sender and the intended recipient to share a key—typically a string of random numbers, often binary—for creating and deciphering secret messages. (See the appendix for an example.) Such a cryptosystem is completely secure, provided that the key's sequence of digits is truly random and that the key, which must be as long as the message, is used only once. Indeed, it is the randomness of the key that wipes out any patterns that could be used by code breakers to crack ciphers. Without a key to decipher it, the cryptogram itself looks like a string of random numbers.

Known as the "onetime pad" system, this scheme was used for German and Soviet diplomatic communications before and during World War II and by Communist spies during the cold war. The name comes from the practice of printing the key in the form of paper pads, each sheet of which would be torn out and destroyed after being used just once.

In one case during World War II, a Soviet spy ring based in Switzerland communicated with Moscow by taking numbers from a book of Swiss trade statistics. Each page had long lists of numbers, and a spy would randomly pick a page, row, and column at which to begin the key. The encoded message itself would include a specially coded passage in a known location that would tell the decoders back in Moscow where in the book to look to find the key for deciphering the message.

Though truly unbreakable, the onetime pad and related systems suffer from a serious practical drawback. The two parties involved in the transaction must initially meet to decide on a key or take the risk of having the key hand-delivered by a courier or transmitted over a telephone line. Either way, the logistics are cumbersome and the exchanges are potentially insecure, especially because keys are lengthy and must be changed regularly to increase security.

Such difficulties have spawned a considerable mathematical effort to invent alternative cryptosystems that may not be unbreakable but nonetheless are sufficiently difficult to crack that the risk is worth taking. In each case, the systems give up randomness for the sake of convenience. Public-key cryptography, for example, requires the use of a pair of complementary keys instead of a single, shared key. One key, which is openly available, is used to encrypt information; the other key, known only to the intended recipient, is used to decipher the message. In other words, what the public key does, the secret key can undo. Moreover, the secret key can't be easily deduced from the public key.

The most popular type of public-key encryption, invented by Ronald Rivest of MIT, Adi Shamir of the Weizmann Institute of Science in Israel, and Leonard Adleman of USC, is known as the RSA cryptosystem. In the RSA scheme, the secret key consists of two prime numbers that are multiplied to create the lengthier public key. Its security rests on the observation that it's much easier to multiply two primes to obtain a large number as the answer than it is to determine the prime factors of a large number. Factoring numbers more than 150 digits long, for instance, presents a formidable computational challenge, but it's relatively easy to identify a pair of 75-digit primes.

Whatever the cryptographic scheme, the issue comes down to using keys that consist of patternless strings of digits—random numbers. Identifying such strings for cryptographic purposes, however, turns out to be a tricky proposition (see chapter 9).

In error correction and encryption, we toy with randomness. We impose regularity on our signals to provide mathematical insurance that their meaning emerges unscathed amid the vagaries of an imperfect material world. We add precise doses of randomness to our signals to disguise our intentions. Lurking in the background, however, is the uncertainty that is built into the physical world—an unpredictability born of physical law. We call it chaos.

7

Complete Chaos

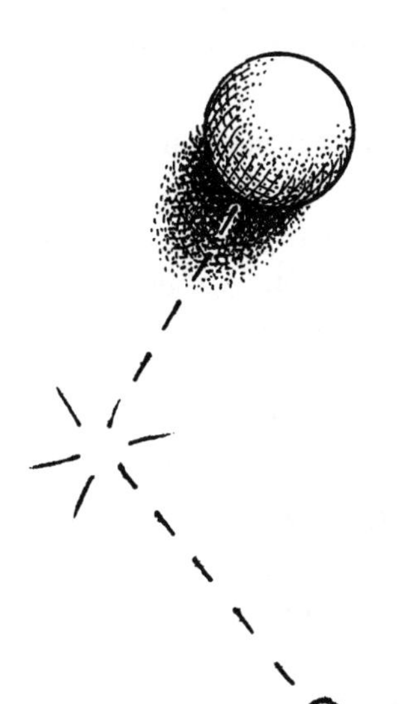

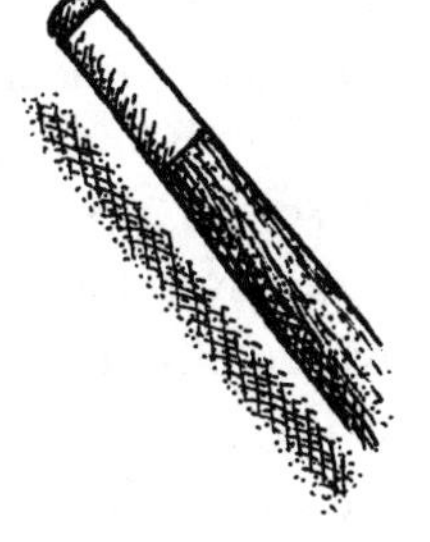

At which the universal host up sent
A shout that tore hell's concave, and beyond
Frighted the reign of Chaos and old Night.

—*John Milton (1608–1674),* Paradise Lost

Cones of smoky light soak into the green felt of a billiard table. The player squints at the gleaming balls scattered across the surface, adjusts his stance, and lowers his billiard cue, aiming the tip directly at one ball. A smooth stroke ending in a sharp tap shoots the ball across the surface. The ball bounces off one cushion, then another before smacking squarely into a second ball.

On a level, rectangular table, a skillful player can return the balls to their original positions, repeat the shot, and obtain the identical result. Even an expert would be hard pressed to accomplish the same feat, however, on a rounded, stadium-shaped table. In stadium billiards, a slight difference in starting point can radically alter a ball's trajectory. The more often the ball rebounds from the curved walls, the less predictable its path becomes. If there were no friction at all, the ball's continuing motion as it rattled around the table would appear quite random.

Unlike rolling dice or tossing coins, billiards is generally considered a game of skill rather than of chance. Yet, under certain conditions, the game can appear extremely chancy. Depending on the billiard table's configuration, a ball's motion can be either predictable or chaotic. The same physical laws determine the outcome in either case, but in the chaotic situation the result is so sensitively dependent on initial conditions that the ball's path quickly becomes unpredictable.

A rectangular billiard table studded with round obstacles, or bumpers, produces the same kind of unpredictability as a stadium-shaped table. In trying to duplicate a shot, a player would find that the second ball's trajectory rapidly diverges from that of the first, with each bounce of the ball off a bumper magnifying any deviation from the first ball's path.

A pinball machine takes advantage of such uncertainty to add elements of suspense and surprise. In the days before electronic extravaganzas equipped with flashing lights, sound effects, fancy bumpers, and flippers, the pinball machine was simply a sloping surface punctuated by an array of pins. A player would insert a coin, pull back a spring-loaded shooter, fire a metal ball to the top of the array, then watch the ball roll down the slope. Every time the ball struck a pin, its path would change, sometimes going to one side and sometimes to the other, eventually ending up at the bottom.

Often installed at drugstores and eating establishments, such pinball machines were a fixture—and an addictive pastime—of the 1930s. Edward N. Lorenz, who is now a meteorologist at the Massachusetts

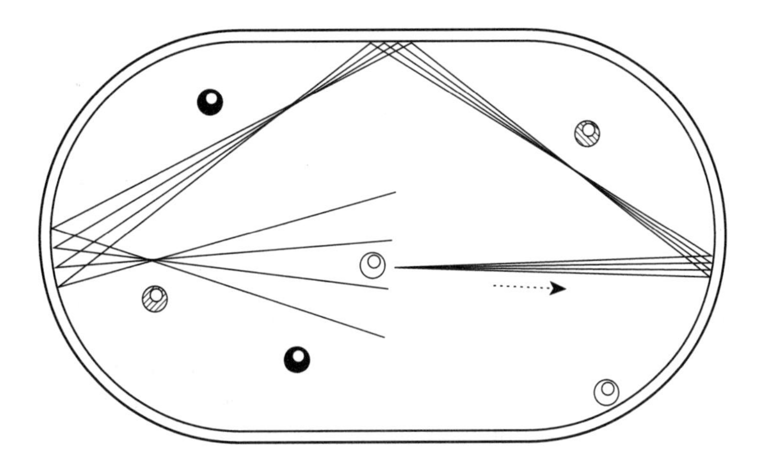

On a stadium-shaped billiard table, balls that start off in slightly different directions follow paths that diverge rapidly with each bounce.

Institute of Technology, has described how installation of the machines affected student life in the college town of Dartmouth, where he was an undergraduate during those years. "Soon many students were occasionally winning, but more often losing, considerable numbers of nickels," he wrote in *The Essence of Chaos.* "Before long the town authorities decided that the machines violated the gambling laws and would have to be removed, but they were eventually persuaded, temporarily at least, that the machines were contests of skill rather than games of chance, and were therefore perfectly legal."

Lorenz's account highlights the curious, tangled relationship between what we consider to be random events—the outcomes of chance—and deterministic occurrences—results that follow strictly from well-defined physical laws. At the heart of this tangle lies the phenomenon we now call chaos.

Amusements and Thrills

Much of the fun of an amusement park ride results from its stomach-churning, mind-jangling unpredictability. The Tilt-A-Whirl, for example, spins its passengers in one direction, then another, sometimes hesitating between forays and sometimes swinging abruptly from one motion to another. The rider never knows exactly what to expect next.

Yet these complicated, surprising movements result from a remarkably simple geometry. A passenger rides in one of seven cars, each mounted near the edge of its own circular platform but free to pivot about the center. The platforms, in turn, move at a constant speed along an undulating circular track that consists of three identical hills separated by valleys, which tilt the platforms in different directions. The movements of the platforms are perfectly regular, but the cars whirl around independently in an irregular manner.

One can find a mathematical equation that approximates the motion of an idealized Tilt-A-Whirl. In essence, the movements of an individual car resemble those of a friction-impaired pendulum hanging from a support that is both rotating and being rocked back and forth while the pendulum swings. Solving the equation determines how a Tilt-A-Whirl car would behave under various conditions.

The mathematical model indicates that when the platforms travel at very low speeds along the track, the cars complete one backward revolution as their platforms go over each hill. In contrast, at high speeds a car swings to its platform's outer edge and stays locked in that position. In both cases, the motion is quite predictable.

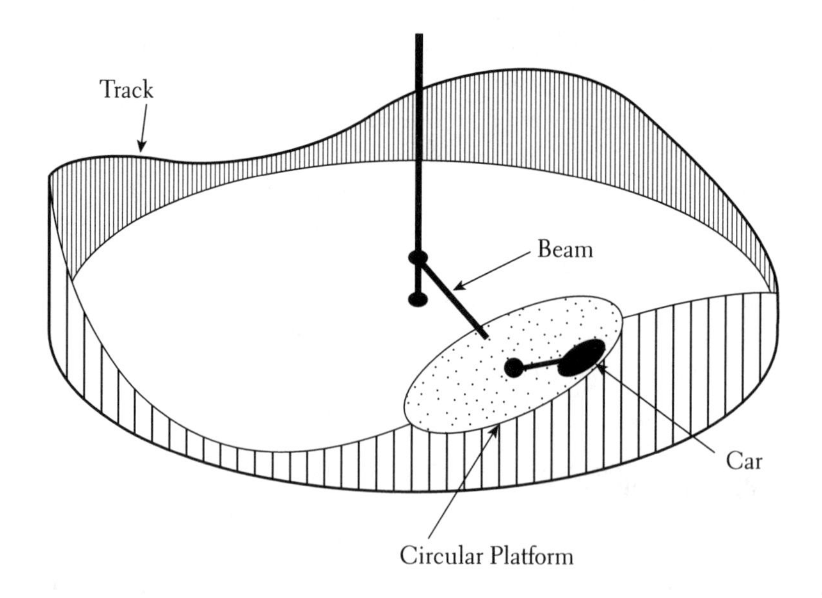

This schematic view of the Tilt-A-Whirl shows a single platform, attached to a beam linked to the center of the mechanism, moving along an undulating circular track. The wheeled car is pivoted from the platform's center.

Chaotic motion occurs at intermediate speeds, close to the 6.5 revolutions per minute at which the ride actually operates. What happens to an individual car is highly dependent upon the weight of its passengers and where they sit. The resulting jumbled mixture of car rotations never repeats itself exactly, which gives the Tilt-A-Whirl its lively and unpredictable character. Indeed, no two trips are ever likely to produce exactly the same thrills and chills. "A walk around an amusement park suggests that several other common rides display chaotic behavior similar to that of the Tilt-A-Whirl," say the physicists Bret M. Huggard of Northern Arizona University and Richard L. Kautz of the National Institute of Standards and Technology, who analyzed the Tilt-A-Whirl's dynamics. Typically, rides that fit this category have cars that are free to rotate or shift back and forth as they follow a fixed track.

The Tilt-A-Whirl first operated in 1926 at an amusement park in White Bear Lake, Minnesota. Most likely, the ride's inventor, Herbert W. Sellner, discovered its unpredictable dynamics not through mathematical analysis but by building one and trying it out. "Ride designers have been fairly adept at finding chaos without appreciating the mathematical underpinning of what they're doing," Kautz notes. Now that may be changing. To fine-tune the thrills, manufacturers are beginning to use mathematical analyses and computer simulations to help build chaotic motion deliberately into future rides.

The amusement park isn't the only setting in which chaotic behavior arises. Observed in a leaking faucet's irregular drip, a laser's erratic light flashes, or a human heart's subtly varying beat, chaos appears in many different situations. To model the dynamics of chaotic systems, mathematicians, scientists, and engineers use equations that describe how the positions and velocities of the systems and their components change over time in response to certain forces.

It's convenient to characterize a system's dynamics by plotting how its position and velocity evolve over time. Each plotted point represents the system's state of motion at a particular instant, and successive points generate a winding line through an imaginary mathematical space representing all possible motions. Different starting points generally initiate different curves.

A simple, repeating motion, like the to-and-fro oscillations of a swinging pendulum, appears as a circle or some other closed curve. Such a plot shows that the system cycles through precisely the same state of motion again and again at regular intervals. More complicated sequences of movements produce tangled paths that wander through space, sometimes never forming a closed loop. That's what Kautz and Huggard found for their equation describing the Tilt-A-Whirl at intermediate platform speeds.

Often, it helps to examine such complicated movements not at every moment but at predetermined, regular intervals. In other words, one starts with a point representing the system's initial state, then waits a given time and plots a second point to give the system's new state, and so on. In the case of a simple pendulum, selecting an interval equal to the time it takes the pendulum to complete one oscillation produces a plot that consists of a single point. Assuming that no friction affects its behavior, the pendulum is always back in its initial state at every repeated glimpse of its motion.

When the motion is chaotic, however, there is no characteristic period. The resulting plot, known as a Poincaré section, shows points scattered across the plane—like bullets puncturing a sheet of paper. In a sense, the system is continually shifting from one unstable periodic motion to another, giving the appearance of great irregularity. Thus, when Kautz and Huggard plotted a set of points representing the velocity and angle of a car at the beginning of each of a hundred thousand tilt cycles, they found that the values never repeated themselves but were scattered in a distinctive swirling pattern confined to a portion of the plane. Although the cars appear to jump randomly from one state of motion to another, these states are themselves not randomly scattered among all the possible types of motion.

So, there's some underlying order in the sustained, irregular behavior characteristic of chaos. The distinctive pattern of points observed in the Poincaré section is known as an attractor. Different equations and different dynamical systems produce different attractors. In a set of equations describing air flows in the atmosphere, for example, Edward Lorenz discovered a distinctive array of points, which looks somewhat like a butterfly with outstretched wings. (This is shown in the figure on the next page.)

For certain equations, such as the one describing the Tilt-A-Whirl for intermediate platform velocities, even slight changes in starting point lead to radically different sequences of points (though the overall shape of the attractor remains roughly the same for each sequence). At the same time, it becomes virtually impossible to predict several steps ahead of time precisely where a particular trajectory will go. Such sensitive dependence on initial conditions stands as a hallmark property of chaos.

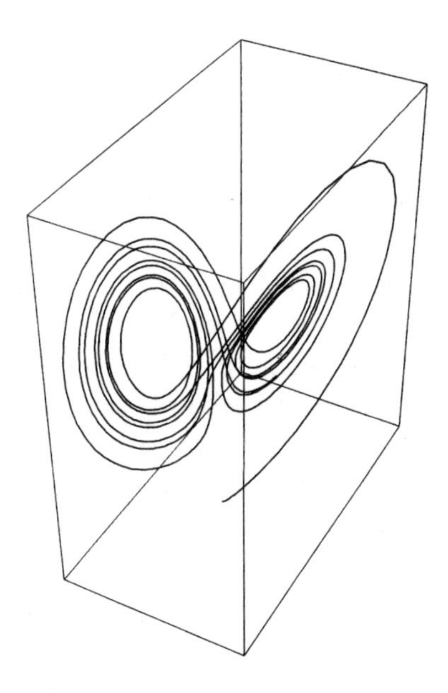

Edward Lorenz's butterfly attractor is one way of representing particular solutions of a set of three equations that serve as a simple model of weather systems. In three dimensions, the resulting pattern consists of two spiral "wings" at an angle to each other. Though the lines look solid, they actually consist of thousands of closely spaced points.

Interestingly, Tilt-A-Whirl fanatics know by experience that they can actually take advantage of this sensitivity. They can affect the motion of a car by throwing their weight from side to side at crucial moments, turning cycles with little or no rotation into thrilling whirls. "Thus, it would seem that aficionados of the Tilt-A-Whirl have known for some time that chaotic systems can be controlled using small perturbations," say Kautz and Huggard.

Scientists and engineers have only recently begun to apply the same principle to the control of chaos in electronic circuits and lasers and to the management of chaos in the heart and the brain. The goal is to stabilize and suppress chaos in some cases and to maintain and enhance it in other situations.

Dance Steps

Initially standing upright like a stiff stalk of grass, a paper-thin metal ribbon begins to bend over. It bows low, then straightens. It bends again, this time making only a quick nod. Then it does a string of low bows, a few brief nods, and another low bow, sometimes throwing in a wiggle or two. Wavering unpredictably, the ribbon's motion follows no apparent pattern.

A researcher then begins adjusting the magnetic field in which the ribbon dances. In a few seconds, the ribbon's chaotic motion settles into a repeating pattern: first a deep bow, then a nod, then another deep bow, a nod, and so on. Instead of abruptly and arbitrarily shifting from one type of motion to another, the ribbon endlessly repeats the same combination of moves.

This laboratory experiment vividly demonstrates that one can suppress chaos in a real-world physical system simply by making small adjustments to one of the parameters governing the system's behavior. The first indication that such control could be attained surfaced in 1990. In a theoretical paper, Edward Ott, Celso Grebogi, and James A. Yorke of the University of Maryland introduced the notion that—just as small disturbances can radically alter a chaotic system's behavior—tiny perturbations can also stabilize such activity.

The success of this strategy for controlling chaos hinges on the fact that the apparent randomness of a chaotic system is really only skin-deep. Beneath the system's chaotic unpredictability lies an intricate but highly ordered structure—a complicated web of interwoven patterns of regular, or periodic, motion. Normally a chaotic system shifts continually from one pattern to another, creating an appearance of randomness.

In controlling chaos, the idea is to lock the system into one particular type of repeating motion. The trick is to exploit the fact that a chaotic system already encompasses an infinite number of unstable, periodic motions, or orbits. That makes it possible to zero in on one particular type of motion, called a periodic orbit, or to switch rapidly from one periodic orbit to another.

Ott and his coworkers successfully demonstrated control of chaos in a numerical experiment on a computer. "Far from being a numerical curiosity that requires experimentally unattainable precision, we believe this method can be widely implemented in a variety of systems, including chemical, biological, optical, electronic, and mechanical systems," they predicted.

The first laboratory demonstration of the control method proposed by Ott, Grebogi, and Yorke occurred a few months later at the Naval Surface Warfare Center in Maryland, where William L. Ditto, Mark L. Spano, and Steven N. Rauseo were experimenting with a most peculiar material—a specially prepared iron alloy that changes its stiffness in accordance with the strength of an external magnetic field.

The metal strip—which resembles a piece of stiff tinsel and is planted in a clamp to keep it upright—stands erect at first. Electromagnetic coils the size of bicycle rims generate a magnetic field along the ribbon's length. Increasing the magnetic field causes the strip to soften so that the ribbon begins to droop under its own weight. Once the external magnetic field reaches a value only a few times larger than Earth's modest field, the ribbon abruptly begins to stiffen again and straighten.

By imposing a slowly oscillating magnetic field, the researchers discovered that they could force the ribbon to droop and straighten in sync with the field. It was like turning a dial to make the ribbon softer or stiffer and letting gravity do the rest. Increasing the field's frequency or strength beyond a certain limit, however, pushed the ribbon into a chaotic regime in which its motion became unpredictable.

By monitoring the ribbon's displacement at the end of each cycle of the oscillating magnetic field, Ditto and his coworkers constructed a kind of map of this irregular motion—very much like the Tilt-A-Whirl's Poincaré section. If the ribbon repeated the same motion at regular intervals, its position would be the same at the end of each cycle. The entire map would consist of a single point. If the ribbon's motion was irregular, however, its position would vary each time, and the resulting map, built up point by point, would show the system's attractor as a swath of scattered dots.

To control the ribbon's motion, the researchers displayed the attractor on a computer screen and selected a point that corresponded to a particular type of periodic motion. Then they waited until the ribbon's movements, indicated by successive dots on the computer screen, happened to wander into a state close to that represented by the point they had selected on the attractor. That close encounter would trigger a simple computer program that calculates how much and in which direction the underlying, steady magnetic field should be adjusted in order to keep the ribbon trapped in the desired periodic motion.

In effect, corralling the ribbon's oscillations is akin to balancing a ball on a saddle. The ball won't mount the saddle's raised front or back, but it can easily roll off at the sides. By continually monitoring the ball for any slight movements to one side or the other, it's possible to counter the tendency and keep the ball in place simply by briefly tilting the saddle in the appropriate direction.

By making slight adjustments about once a second to the steady vertical magnetic field acting on the ribbon, Ditto and his team could maintain the ribbon's regular motion for as long as they wished. As soon as they relinquished control, the ribbon would return to its chaotic dance. They could then reestablish control, bringing the ribbon back to the same periodic motion it had before, or, just as easily, putting it into a different type of regular motion.

Surprisingly, it turned out that one didn't even need to know the equations that describe the behavior of a particular system in order to make the control technique work. It was sufficient to observe the system long enough to map its chaotic attractor and to determine by experiment a few crucial quantities necessary for establishing control. All that was needed, really, was a saddle whose shape characterized the system's dynamics, and this could be obtained simply by observing the system.

It wasn't long before research teams found that they could use similar approaches to subdue the fluctuating output of high-power lasers and the erratic electrical activity of certain electronic circuits. These successes opened up new possibilities for scientists and engineers, who had previously considered a chaotic system's extreme sensitivity to initial conditions something to be avoided. "This characteristic is often regarded as an annoyance, yet it provides us with an extremely useful capability without a counterpart in nonchaotic systems," Ott, Grebogi, and Yorke argue. "In particular, the future state of a chaotic system can be substantially altered by a tiny perturbation. If we can accurately sense the state of the system and intelligently perturb it, this presents us with the possibility of rapidly directing the system to a desired state."

One can envision, for example, a hypothetical set of chemicals that, when mixed in different concentrations, reacts to produce dyes of different colors. Chemical engineers might design a factory for producing various dyes by building separate chemical reactors, one for each color. But by introducing and controlling chaos, they could get away with building a single reactor. If they adjusted the ingredient concentrations so that the system's products would shift chaotically from one color to another, then a simple control scheme involving tiny adjustments in the concentration of one ingredient would suffice to stabilize the reaction so that it produced only one color. The same control system would also allow them to shift dye production rapidly from one color to another.

In fact, the sensitivity to initial conditions that characterizes chaos could offer considerable advantages, allowing system designers greater flexibility in their choice of materials and architectures and making possible multipurpose systems that adapt quickly to changing needs. Just as gearing allows you to operate a heavy truck with a few fingers on the steering wheel, chaotic dynamics potentially permits a similar ease of control in a variety of situations.

The most intriguing possibilities for control may involve biological systems. Research teams have already managed to stabilize the irregular heartbeats of a mass of heart cells and to turn on and off seemingly random patterns in the electrical activity of neurons. Such successes suggest that biological systems may use chaos and the richness of chaotic dynamics to adjust their behavior on the fly. Multipurpose flexibility is essential to higher life-forms, and chaos may be a necessary ingredient in the regulation of the brain. Controlled chaos in the brain's juggling of electrical signals, for example, may play a role in the human ability to produce strings of different speech sounds quickly enough to carry on an intelligible conversation.

Riddled Basins

"Sunny, bright, milder. High 58. Low 42. Wind 4–8 miles per hour."

The morning's weather forecast gives us a rough idea of what to expect during the rest of day. But same-day weather predictions are sometimes quite wrong, failing to foretell the sudden afternoon shower or the sweltering humidity brought on by a stalled air mass. Nonetheless, meteorologists predicting the weather a day or so in advance are correct far more often than they would be if they were simply guessing. Weather predictions several days into the future, however, are considerably more unreliable. In fact, there isn't much point in looking ahead more than a week. Although all weather systems are governed by physical laws, which should allow meteorologists to make predictions, the churnings of the atmosphere and the oceans are extremely complicated. Moreover, it's quite possible that the underlying dynamics of Earth's air and water flows is chaotic, which would limit how far into the future it's even possible to make reliable predictions.

Think of the daily fluctuations in temperature, precipitation, and wind that constitute weather as the points of a Poincaré section. And think of the pattern that these points make, forming an attractor, as the climate, which is relatively stable. We know, for instance, that on the average, summers will be warmer than winters; that temperatures rarely exceed 120 degrees Fahrenheit; and that wind speeds seldom approach 150 miles per hour. The geometry of the climate attractor suggests that there are inherent limits to the variability of global weather.

Suppose, however, that the climate itself could change abruptly from day to day. You might wake up one morning in the tropics and the next morning in the throes of an ice age. Luckily, barring a cataclysmic collision between Earth and a massive asteroid, there's no evidence that such drastic, rapid shifts in climate could occur. Nonetheless, mathematicians have pinpointed certain features of equations, including some of those used to describe physical phenomena such as fluid flow, that lead to an extreme kind of unpredictability. It's as if not only the weather but also the climate could vary from one day to the next.

The first person to identify this bizarre type of behavior in an equation was James C. Alexander, a mathematician at the University of Maryland. At a dynamics meeting several years ago, he encountered a puzzling image—just two crossed lines projected on a screen. The emergence of this X from a set of simple equations being studied by a researcher didn't seen to fit with what Alexander knew of the mathematical behavior of dynamical systems. "It was a complete mystery," he recalls. "In fact, while everybody else was having a nice dinner that night, I was scribbling on the back of an envelope trying to figure out what was going on."

Mathematicians are particularly interested in what happens to sequences of points, or trajectories, representing a system's dynamics at given time intervals (as depicted in a Poincaré section). In some cases, certain collections of starting points lead to the same endpoint or to a particular group of endpoints to generate the system's attractor. The area covered by starting points that eventually arrive at an attractor is called a basin.

Despite an extreme sensitivity to initial conditions, chaotic trajectories in a given dynamical system still generally end up on an attractor that has a particular geometry—albeit one, like Lorenz's butterfly attractor, that often looks extremely convoluted and complicated. Such an attractor, however, normally doesn't include crossings or sharp corners—and certainly not a crisp, unambiguous X.

By taking a close look at the equations that originally piqued his curiosity, Alexander came to realize that the corresponding dynamical system apparently has three attractors. The attractors in this case are incredibly simple. They are just line segments, which intersect to form the outline of an equilateral triangle. The researcher's X was merely one of the three places where these line segments cross. Alexander found that any starting point already on one of the lines would follow a trajectory that skips erratically back and forth along the line without ever hopping off. This behavior furnishes evidence that the lines are chaotic attractors.

What's surprising is the behavior of starting points chosen from areas close to the lines or inside the triangle. If one starting point follows a trajectory leading to one of the line-segment attractors, then another starting point only a tiny distance away may end up on a different attractor. There's no way of predicting which attractor a given starting point will land on.

Normally, if you start at some point and then start at a slightly different point, you expect to come down on the same attractor. In the situation that Alexander had uncovered, if a point goes to one attractor, then arbitrarily close to it, there are points that go to another attractor. In other words, the system is extremely sensitive to initial conditions not only for determining where on an attractor a given point will land (as in chaos) but also for deciding which attractor it will fall on. Hence, instead of having a single basin of attraction, this dynamical system has three thoroughly intermingled basins. In effect, the slightest change in initial conditions can completely alter the system's type of activity (see color plate 5).

Computer-generated images in which points in a plane are colored according to which attractor they land on clearly shows the intricacy of the meshed basins belonging to the attractors. (See the appendix for an example of how to compute such a portrait.) The images suggest the idea of a "riddled" basin. Each basin is shot full of holes. No matter where you are, if you step infinitesimally to one side you could fall into one of the holes, which means you could end up at a completely different attractor.

Alexander's pursuit of this curiously aberrant behavior did more than confirm the presence of the unlikely X. It eventually unveiled a weird mathematical realm even stranger and, in some sense, wilder and more unpredictable than that found in dynamical systems generally described as chaotic.

How common is the phenomenon of intermingled, riddled basins? Could this type of behavior occur in the mathematics that describes a physical situation?

No one has all the details yet, but soon after Alexander's discovery, James Yorke identified a second set of equations that displays similar sensitivities. These equations are sometimes used in mathematical biology to describe fluctuations in the populations of two competing species. A little later, the physicists Edward Ott and John Sommerer found evidence of the same effect in an equation representing the motion that results from a particle traversing a force field with a particular geometry. The particle, which is periodically jolted as it moves, also experiences a frictional force that depends on its velocity.

In this situation, the particle can settle into a chaotic type of motion, or it can be forced either toward or away from infinity. Which one of the three courses the particle takes depends on its starting position and its velocity. No randomness is built into the model, yet the final state of the system cannot be predicted with certainty if there is any error (no matter how small) in the measurement of the initial conditions. It isn't even possible to make a rough guess about the system's ultimate fate.

Thus, simple physical systems—governed by nothing more than Newton's laws of motion—can display the pathology that Alexander initially found in a more theoretical, abstract setting. The ease with which Ott and Sommerer found their example and the fact that there is nothing particularly special about the chosen equations of motion suggest that "riddled" systems may be relatively common—albeit not as ubiquitous as chaotic systems. As a result, even qualitative reproducibility in simple classical systems can't be taken for granted. "Clearly, in a philosophical sense, Nature isn't done throwing curves at us," Alexander says.

The discovery of such effects has intriguing implications for science. The scientific method hinges on the ability of researchers to perform reproducible experiments. What one scientist has measured, another should be able to replicate under the same conditions. But what can one do when the slightest error in reproducing an experiment's initial conditions can lead to vastly different outcomes?

Indeed, little is more disconcerting and frustrating to a scientist than discovering that supposedly identical experiments somehow produce radically different results. Such perplexing, discrepant data would typically end up in a wastebasket rather than in a journal article. "You have a deterministic system, yet you lose experimental replicability," Alexander says. "You're always going to have little errors, and such small changes in initial conditions may lead to completely different long-term behavior."

What's remarkable about this work is that for the equations themselves, derived from applications in physics and biology, nothing has changed. Strange behavior has always been there, but no one had previously thought to look for it.

Troubling Complexities

Erratic change is a common feature of everyday life, whether in the zigzagging flight of a rapidly deflating balloon across a room, the fluctuations of the stock market, the wavering passage of a falling leaf, or the wispy trails of smoke from a smoldering fire. In many cases, irregular movements represent the net effect of numerous separate events. The resulting effects may be considered essentially random (see chapter 8). However, simple systems can also have complex, unpredictable outputs, as demonstrated in the motion of the Tilt-A-Whirl or the wildly unpredictable gyrations of a double pendulum.

A single pendulum consisting of a rod pivoted at one end—like the timekeeper of a grandfather clock—simply swings back and forth. The addition of a second rod, pivoted from the bottom of the first, greatly increases the motion's complexity. Such a double pendulum can swing placidly for a while, then abruptly shift to large gyrations and wide arcs, with the two sections of the pendulum sometimes moving in concert and sometimes independently. All the while, the lower rod performs maneuvers worthy of a trapeze artist or a champion gymnast on the high bars.

In studies of the dynamics of many physical and biological systems, researchers face the dilemma of determining from experimental data whether observed variations represent random fluctuations (or noise), the chaotic state of a deterministic system, or some combination of the two. When they can demonstrate that a system is chaotic rather than random, they will have a better chance of developing a strategy to understand and control its irregular behavior.

Consider, for example, the human heartbeat. An electrocardiogram typically shows a series of spikes, each one marking an electrical pulse

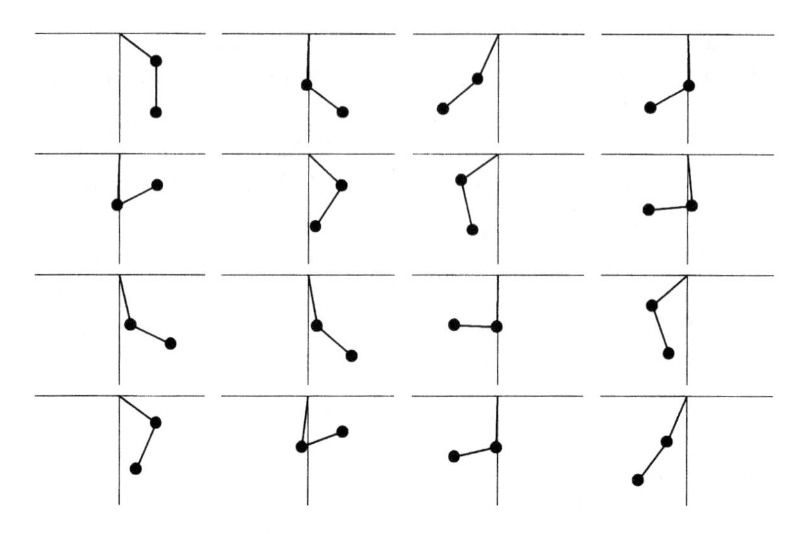

A double pendulum, in which one rod is pivoted from the bottom of the first, can show a wide range of unpredictable, capricious movements.

associated with a beat. Cardiologists need only look at the rate and the pattern of spikes to distinguish between a healthy heart and an ailing one. In extreme cases, a normally beating heart can suddenly shift from a regular rhythm to an erratic, deadly shivering known as cardiac fibrillation.

One can look for both pattern and chaos among the signals in a variety of ways. For example, Ary L. Goldberger of Boston University and his colleagues have studied the intervals between heartbeats, in effect turning the electrical activity of the heart into a series of numbers representing the length of time between successive beats. When the heart is beating normally, the intervals between beats appear to fluctuate in a complex, apparently erratic manner. Though the heartbeat is quite regular, the prevalence of slight differences in the timing of beats becomes evident in the data.

Some of these fluctuations can be attributed to the combined effects of breathing cycles and other inputs from the rest of the body. Goldberger and his team have focused on beat-to-beat fluctuations over very long periods, covering up to a hundred thousand beats over the course of a day. The results of their statistical analyses indicate that a healthy heart shows more complicated fluctuations than a diseased heart. While beat-to-beat intervals in a diseased heart rise and fall roughly according to the toss of a coin, those of a healthy heart also exhibit subtle, complex, long-range patterns.

Goldberger argues that a loss of complexity in the heartbeat rhythm signals trouble. "As the system breaks down . . . it loses its variability," he says. That variability is what gives a living, dynamic system, such as the heart, the robustness it requires to cope with change. So, by examining not just a standard electrocardiogram but also the length of the intervals between heartbeats over long periods, cardiologists may be able to identify patients who are at high risk of sudden death from heart disease.

Nonetheless, the cardiovascular system is complicated, and its state can be measured in many different ways. It's not always clear what the data show. Indeed, given a series of measurements strung out in time, characterizing the numbers is rarely straightforward, and the statistical and mathematical methods that researchers use to perform this analysis all have strengths and weaknesses in the kinds of patterns—if any—they are able to locate.

Paul Rapp, a physiologist at the Medical College of Pennsylvania, once checked the efficacy of a variety of tests commonly used to identify and characterize chaotic systems. He applied the tests not to biomedical data but to a set of random numbers. The numbers were adjusted slightly to reflect the kind of filtering typically imposed on experimental measurements to remove extraneous noise when signals pass from sensor to amplifier to computer. Five widely used tests failed to identify correctly the underlying randomness in this artificial data set.

According to Rapp, researchers often use such tests blindly, assuming that the tests unambiguously point to chaotic behavior. His experiment "produced vivid examples of how the casual application of the methods of dynamical analysis produce fallacious conclusions," Rapp notes. "By implication, these results suggest that many of the putative identifications of chaotic behavior, particularly in biological data, are almost certainly spurious."

Published claims of chaotic behavior, particularly in biological systems, simply can't be counted on. But all is not lost. Improved tests are in the works, and researchers have already proposed alternative ways of distinguishing between deterministic and random dynamics, given data in the form of a sequence of observations made at definite time intervals. Sophisticated tests, heedfully applied, can provide useful insights even when applied to "noisy" biological data.

One promising approach involves a measure of irregularity called the approximate entropy. Invented by Steve Pincus, a freelance mathematician in Connecticut, and developed in cooperation with Burton H.

Singer of Princeton University, this measure characterizes the randomness of a sequence of numbers. Because the method does not depend on any assumptions about the process involved in generating a sequence of numbers, it can be applied to biological and medical data and to physical measurements, such as the number of alpha particles emitted by a radioactive element in specified time intervals.

Suppose the data are expressed as a string of binary digits. The idea is to determine how often each of eight blocks of three consecutive digits—000, 001, 010, 011, 100, 101, 110, and 111—comes up in a given string. The maximum possible irregularity occurs when all eight blocks appear equally often.

The technique can even be applied to the digits of such irrational numbers as π, the square root of 2, and the square root of 3 (see chapter 10). Given the first 280,000 binary digits of these three numbers, π turns out to be the most irregular. Pincus has even looked at stock market performance, as measured by Standard and Poor's index of five hundred stocks. His calculations show that fluctuations in the index's value are generally quite far from being completely irregular, or random (see chapter 8). One striking exception occurred during the two-week period immediately preceding the stock market crash of 1987, when the approximate entropy indicated nearly complete irregularity. That change flagged the incipient collapse.

Chaos itself is also just one of a wide range of behaviors possible in nonlinear dynamics. Scientists have traditionally focused on systems that could be modeled by linear equations. With linear relationships, a slight change in a system's starting point means only a slight difference in its final state. By way of contrast, a nonlinear system offers the possibility that a small change can cause a considerable difference in the final result.

Faced with a menagerie of puzzling time-dependent behavior in biomedical processes ranging from erratic heartbeats to brain seizures, researchers are looking for a framework within which to develop an understanding of diverse patterns. Their eternal hope is that the underlying agents responsible for the erratic behavior are not as complicated as the observed activity.

The mathematics of nonlinear dynamics, including chaos, offers a useful theoretical framework for understanding complex rhythms in carefully controlled experimental situations. Unfortunately, successes in studying comparatively simple systems are not easily extended to the difficult clinical problems that confront the practicing cardiologist or neurophysiologist.

In 1972, Edward Lorenz presented a paper describing some of his work on chaotic systems at a meeting in Washington, D.C. The paper had the provocative title "Does the Flap of a Butterfly's Wings in Brazil Set Off a Tornado in Texas?" The image of the nearly negligible actions of a single butterfly triggering a storm somewhere else in the world has become a common illustration of the sensitive dependence on initial conditions that characterizes chaos, and it has appeared in a wide variety of contexts. Lorenz, however, originally had something a little different in mind, and he didn't specifically answer the question posed in the title of his paper. He noted that if a single flap could lead to a tornado that would not otherwise have formed, it could equally well prevent a tornado that would otherwise have formed.

Moreover, the world is full of butterflies. A single flap from one butterfly might have no more effect on the weather than a flap from any other butterfly's wings, not to mention the activities of other species. More often than not, one disturbance balances out the others. Nonetheless, chaos exists. It looks random, but it can be controlled. Indeed, sensitive dependence on initial conditions comes into play in many situations, from stadium billiards and amusement park rides to heartbeat rhythms and electronic circuitry. The trick is to isolate these instances from other phenomena in which many independent agents are at work and the results are truly random.

Still, there are ambiguities. Lorenz asks the question: "Are all irregularities, except those produced by intelligent behavior, chaotic rather than random?" He argues in the affirmative, citing the example of an autumn leaf fluttering to the ground. The leaf's unpredictable movements are guided by the breeze, which in turn is a small part of the global weather system, which behaves with the irregularity expected of a chaotic system.

Similarly, the devices that we use as randomizers—dice, coins, roulette wheels, or decks of cards—also rely for their simulation of chance on a sensitive dependence on initial conditions (see chapter 1) and, hence, on our inability to predict precisely the outcome of a given toss, roll, spin, or shuffle. "I suspect that many other seemingly random phenomena that do not depend primarily on animate activity for their apparent randomness can be analyzed in a similar way," Lorenz concludes. At the same time, he rejects the notion of a strict determinism that says the future course of the universe, including all animate activity within it, is preordained. "We must then wholeheartedly believe in free will," he declares. "If free will is a reality, we shall have the made the correct choice. If it is not, we shall still not have made an incorrect choice, because we shall not have made any choice at all, not having a free will to do so."

Chaos allows us to give up the idea of a clockwork universe set into motion at creation and running relentlessly into the future along its prescribed track. Determinism and predictability are not synonymous. As the advances of mathematics and science over the centuries have demonstrated, we can probe the universe and understand significant parts of it. We are not helpless in the face of chaos and randomness, though we often have to give up absolute certainty, whether in tracking the gyrations of a double pendulum or taking a walk of random steps across a fractured landscape.

8

Trails of a Wanderer

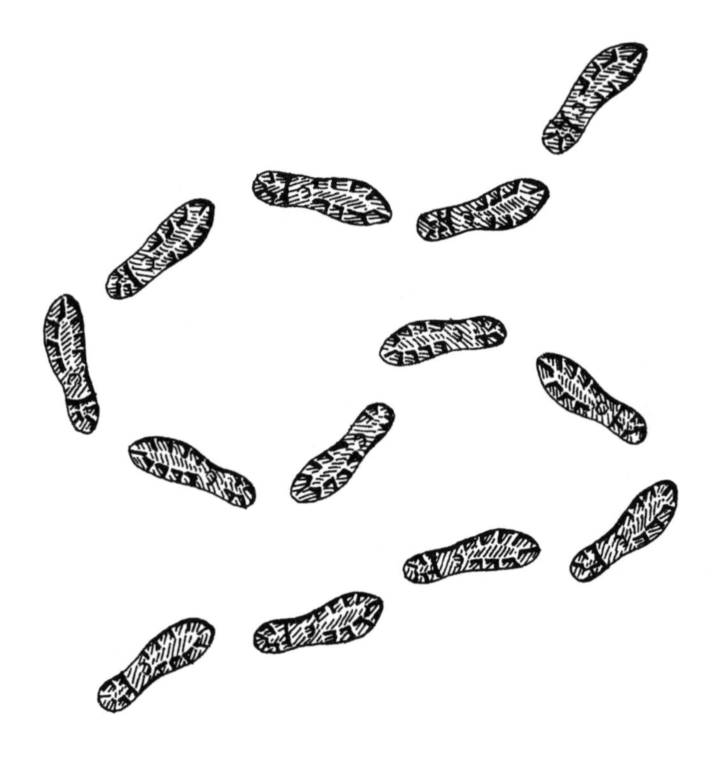

He calmly rode on, leaving it to his horse's discretion to go which way it pleased, firmly believing that in this consisted the very essence of adventures.

—*Miguel Cervantes (1547–1616),* Don Quixote

A sharp knife bites through the thick skin of a clove of garlic, exposing the bulb's pale, layered interior to the air. Within seconds, a pungent, penetrating odor mixes with and quickly overwhelms the buttery aroma of a roasted turkey, the yeasty scent of freshly baked dinner rolls, and the thickly sweet smell of fried onions.

The distinctive tang of freshly chopped or crushed garlic arises from the rupture of cell membranes, which allows the huge protein molecules of an enzyme called allinase to reach and capture tiny molecules of a chemical compound known as alliin. The enzyme stretches and twists the trapped molecules, snapping the chemical bonds and rearranging the atoms of alliin to create the unstable compound allicin. Allicin, in turn, readily transforms itself into diallyl disulfide, which is responsible for garlic's distinctive odor.

As the odor-bearing molecules of diallyl disulfide seep out of the mashed garlic into the air, they begin to drift randomly, jostled by oxygen, nitrogen, and other airborne molecules and transported by wavering currents. Some soon reach the chef's nose. The wandering molecules travel along nasal passages to the back of the nose to reach a delicate sheet of moist, mucus-bathed tissue, where a large number of nerve cells cast their hairy, odor-sensing nets. When a receptor strand snags a molecule, it triggers a series of actions that generates an electric signal, which travels to the brain, and the chef immediately recognizes the smell of garlic.

Our capacity to smell depends on the passage of specific molecules from the source to the sensors in the nose. In contrast, we see and hear because waves of energy, whether electromagnetic oscillations or vibrations of the air, carry the signals. No molecules or atoms actually make the journey from firefly and drum to eye and ear. In the realm of smell, we deal with a statistical process—the effect of thousands upon thousands of random steps as molecules venture into the rough-and-tumble domain of incessant collisions and follow haphazard paths to the nose. One can imagine such journeys as random walks, with molecules traveling in straight lines until they collide. With each collision, they rebound like billiard balls and start off in a new direction—not very different from the haphazard steps that a blindfolded person might take if he or she were walking in a strange landscape studded with obstacles.

Random movements play a significant role in a wide variety of natural phenomena. Hence, the mathematics of random walks is a key element in modeling and understanding not only the jiggling of molecules but also the possible configurations of a polymer's long molecular chain, the flight paths of such foraging birds as the wandering albatross, and even the erratic fluctuations of stock market prices.

Quivering Particles

In 1828, the botanist Robert Brown (1773–1858) published a pamphlet titled "A Brief Account of Microscopical Observations on the Particles Contained in the Pollen of Plants." Brown had traveled widely, exploring Australia and Tasmania before returning to London with specimens of more than four thousand plant species for identification. In the course of establishing a major botanical collection at the National Museum, he had looked closely at pollen grains obtained from many of the collected plants. He noted that, in some cases, "the membrane of the grain of pollen is so transparent that the motion of the . . . particles within the grain was distinctly visible." The suspended granules were only a few micrometers in size, less than one-tenth of a typical pollen grain's width. Observed under a microscope, they appeared to be in continuous, erratic motion.

The Dutch physician and plant physiologist Jan Ingenhousz (1730–1799), who is credited with discovering photosynthesis, had reported something similar in 1785 when he looked closely at powdered charcoal sprinkled on an alcohol surface. Brown's experiments with granules extracted from crushed pollen grains, soot particles, and fragments of other materials suspended in water revealed the same type of unceasing, quivering movement. Some scientists were quick to attribute the effect to minute, heat-induced currents in the liquid surrounding the particles or to some obscure chemical action between the solid particles and the liquid. However, the observation that the movements of two neighboring particles appeared to be quite uncorrelated helped rule out currents in the fluid as a cause.

Other characteristics of Brownian motion were just as intriguing. For example, a given particle appeared equally likely to go in any direction, and its past motion seemed to have no bearing on its future course. Particles also traveled faster at higher temperatures, and smaller particles moved more quickly than larger ones. Most striking of all, the jerky movements never stopped. To explain such observations, some scientists boldly ascribed the phenomenon to molecular movements within the liquid. They pictured the liquid as being composed of tiny molecules whizzing about at high speeds and colliding with each other. When molecules bumped into particles, they would give the particles random pushes.

Although some researchers hailed Brownian motion as visible proof that matter is made up of atoms and molecules, the question of the true structure of matter was a contentious issue. At the beginning of the twentieth century, for instance, the prominent German physical chemist Wilhelm Ostwald (1853–1932) still regarded atoms as merely "a hypothetical conception that affords a very convenient picture" of matter.

In 1905, Albert Einstein provided an elegant explanation of how tiny, randomly moving molecules could budge particles large enough to be observable under a microscope. In his paper "On the Motion of Small Particles Suspended in a Stationary Liquid According to the Molecular Kinetic Theory of Heat," Einstein used statistical methods to show that a suspended particle would get shoved in different directions by the combined effect of the modest impact of many molecules. For particles smaller than about twenty micrometers in diameter, the impact would generally fail to average out equally on all sides, giving the suspended particle a net shove in some direction.

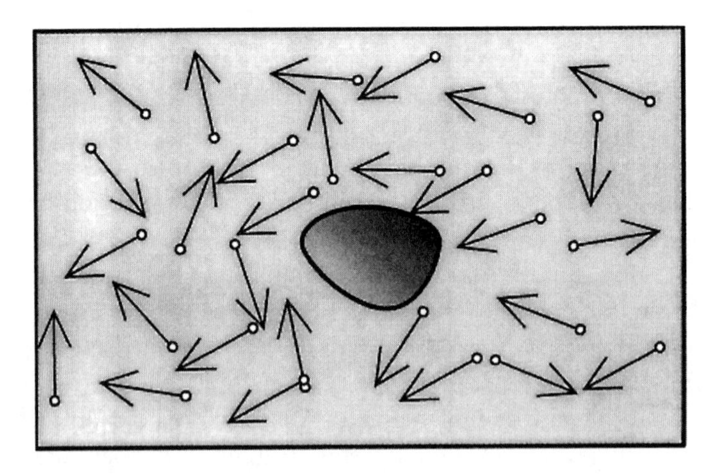

The erratic jiggling of a microscopic particle suspended in water results from the uneven distribution of impacts by water molecules at any given moment.

Interestingly, Einstein had not been aware of the experimental work on Brownian motion. His paper came about because he had begun to consider what effects might follow from the existence of atoms and molecules moving at high speeds that depended on the temperature. His main goal, Einstein later wrote, was "to find facts which would guarantee as much as possible the existence of atoms of definite size." But, he continued, "in the midst of this, I discovered that, according to atomistic theory, there would have to be a movement of suspended microscopic particles open to observation, without knowing that observations concerning the Brownian motion were already long familiar."

A few years later, the French physicist and chemist Jean-Baptiste Perrin (1870–1942) confirmed experimentally some of Einstein's key theoretical predictions. In particular, Perrin and his students were able to track the movements of nearly spherical Brownian particles, which they recorded every thirty seconds and plotted on sheets of paper. Armed with these data, the researchers then used a formula derived by Einstein to determine the number of molecules present in a given volume of fluid.

The experiments gave Perrin a sense of the complexity of the path followed by a Brownian particle. His plots showed a highly irregular track, yet they gave "only a very meager idea of the extraordinary discontinuity of the actual trajectory," Perrin noted. If the researchers could have increased the resolving power of their microscope to detect the effects of bombardment by progressively smaller clusters of molecules, they would have found that parts of a path that initially appeared straight would themselves have had a jagged and irregular structure.

In fact, Brownian motion isn't the only place where such self-similar patterns occur. "Consider, for instance, one of the white flakes that are obtained by salting a solution of soap," Perrin wrote in 1906. "At a distance, its contour may appear sharply defined, but as we draw nearer, its sharpness disappears. . . . The use of a magnifying glass or microscope leaves us just as uncertain, for fresh irregularities appear every time we increase the magnification, and we never succeed in getting a sharp, smooth impression, as given, for example, by a steel ball." Nowadays such patterns, in which the magnified structure looks similar to and is just as complicated as the overall structure, are known as fractals (see chapter 5). And the paths of Brownian particles can be modeled mathematically as random walks.

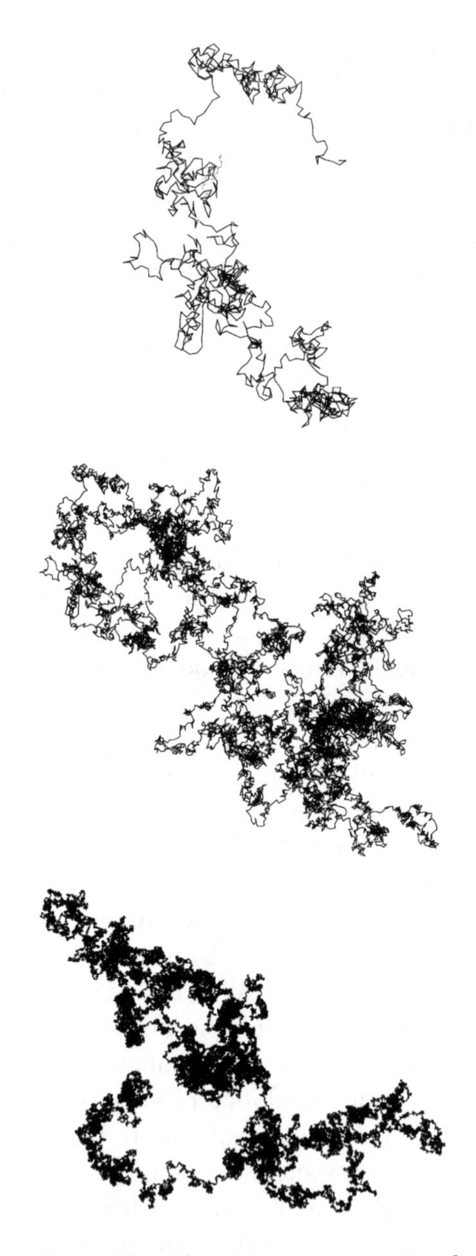

As the number of steps in a computer simulation of Brownian motion increases from one thousand (top) to ten thousand (middle) to a hundred thousand (bottom), the same overall pattern of erratic movement persists, though on increasingly larger scales. (G. M. Viswanathan)

Walking Wild

Though physicists began looking at random processes involving large numbers of colliding molecules near the start of the twentieth century, it wasn't until 1920 that mathematicians began to develop a convincing mathematical model of Brownian motion, starting with the work of Norbert Wiener. At the heart of the mathematics was the difficult problem of making precise mathematical sense of the notion of a particle moving at random.

A Brownian particle suspended in a liquid knows neither when the next shove will occur nor in which direction and how forcefully it will be propelled. Its displacement at any given moment is independent of its past history. These characteristics put Brownian motion in the category of a Markov process, named for the Russian mathematician Andrey A. Markov (see chapter 1). One of the simplest examples of such a process is a one-dimensional random walk, in which a "walker" is confined to a long, narrow path and moves forward or backward according to the results of repeatedly tossing a coin. The walker takes a step in one direction if the outcome is heads and in the opposite direction if the outcome is tails.

For a walk along an infinite track, one can calculate a walker's long-term behavior. The resulting trail wanders back and forth along the track, and the probability of the wanderer's being a certain distance away from the starting point after taking a given number of steps is defined by a bell-shaped curve known as a Gaussian distribution. The larger the number of steps, the wider would be the curve. Indeed, the expected average distance of the walker from the starting point after a certain number of equal steps is simply the length of the step times the square root of the number of steps. For infinitely many coin tosses, a random walk confined to a line corresponds to one-dimensional Brownian motion.

One consequence of this type of erratic movement back and forth along a line is that a random walker is certain to return to the origin (or to any other particular position on the track)—eventually. This might sound like a good strategy for someone who's lost on a tightrope: Just take steps at random and you'll end up anywhere you want to be. But it might take much longer than a lifetime to get there.

It's straightforward to extend the random-walk model to two dimensions: Take steps to the east, west, north, or south, randomly choosing each direction with equal probability (perhaps by using a

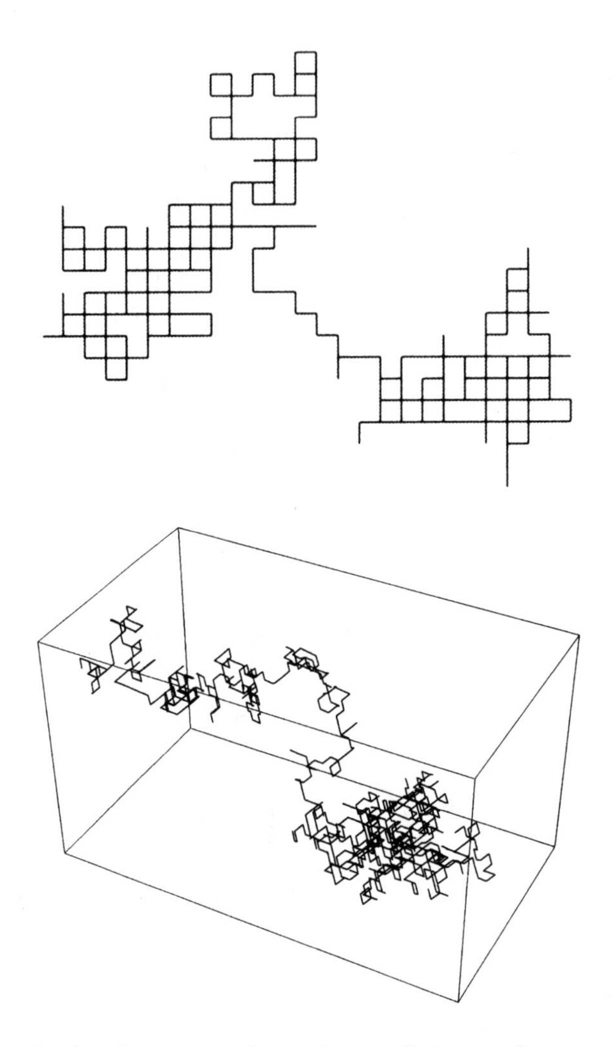

In this illustration of a random walk in two dimensions, a walker starts off from a point on the left-hand side of a checkerboard grid, taking random steps to the east, north, west, or south (top). A three-dimensional random walk typically ranges much farther afield (bottom).

tetrahedral die). One can imagine this walk going from vertex to vertex on an infinite checkerboard lattice. If such a walk continues for an arbitrarily long time, the walker is certain to touch every vertex, including a return visit to its starting point. The fact that returning to the origin is guaranteed in one and two dimensions suggests that there will be infinitely many returns. Once a walk gets back to the origin, it's like starting from scratch, and there has to be a second return, then a third, and so on. Similarly, such a random walk will visit every point infinitely many times.

Things are a little different in three dimensions. A walker can go up and down as well as in each of the four compass directions, so a standard die serves as a suitable randomizer to determine the movements. This time, however, even if a walker takes infinitely many steps, the probability of returning to the origin is only about .34. There's so much room available in three dimensions that it becomes considerably harder for a walker to find the way back to the starting point by chance (see color plate 6).

Indeed, this mathematical result affords an important lesson for anyone who is lost in space. Unless you happen to make it home again within your first few steps, you're likely to end up lost forever. No amount of aimless wandering will get you back after such a start. There are simply too many ways to go wrong.

What Wiener did in the 1920s was to transform a random walk, which is made up of discrete steps, into a suitable mathematical model of Brownian motion. He did it by making the steps or time intervals between steps infinitesimally small. In Wiener's rigorous approach, once the position of a particle is established at the start, its position at any later time is governed by a Gaussian curve, as it is in Einstein's physical model of Brownian motion.

Wiener also proved that although the path of a Brownian particle is continuous, at no point is it smooth. Such a curious, incredibly jagged mathematical curve actually makes physical sense because a particle in Brownian motion can't jump instantaneously from one position to another, so its path must be continuous. At the same time, as Perrin noted, erratic changes in direction appear to take place constantly, so one might expect the path to consist entirely of sharp corners. In fact, a two-dimensional Brownian trajectory wiggles so much that it ends up filling the entire area over which the motion occurs.

With Brownian movement put on a solid mathematical footing by Wiener and others who followed his lead, such abstract formulations began to play a significant role in the creation of models of random phenomena. They were used to represent the diffusion of heat through a metal, the spread of flu epidemics via the random walks of microbes, the structure of background noise and static affecting radio signals, the transport of perfume molecules from source to nose, and even the spread of rumors.

Wiener himself applied his model of Brownian movement to the problem of electronic interference, which disturbs the transmission of radio signals and causes the type of static heard on an AM radio between stations. Because this static has a strongly random character, Wiener was able to use the mathematics of Brownian motion to design an electronic filter to separate the signal from the background noise. Applied to the development of radar during World War II, his results were long kept a military secret.

It is interesting to note that applications of Brownian motion often began with the study of biological processes. The law of diffusion and models of the spread of heat in a material, for instance, initially arose from studies of heat generation in muscles.

One of the first major uses of probabilistic methods in computers was in calculating the random walks of neutrons through different materials—a crucial issue in the design of nuclear weapons and atomic power plants after World War II (see chapter 9). Physicists have applied similar techniques to show that a particle of light near the sun's center takes about fifty centuries to stroll its random walk to the surface before finally escaping the sun and speeding to Earth in about eight minutes.

Mathematicians and scientists have also extended the random-walk and Brownian-motion models to encompass other types of phenomena. The long molecular chain of a polymer floating in a solvent, for example, resembles a miniature, truncated version of the path of a particle undergoing Brownian motion. In other words, one can imagine the chain's small molecular units (called monomers) as points and the bonds between the units as the steps of a three-dimensional random walk.

However, to account for the fact that no two monomers can occupy the same region in space, the random walk has to be modified. A more realistic model is the self-avoiding random walk, which is a path that doesn't intersect itself. Such walks spread out much faster and tend to cover a larger area or volume than their standard counterparts.

Using a self-avoiding random-walk model, polymer scientists can tackle such questions as: How many possible configurations can a long polymer chain adopt? What is the typical distance separating a polymer's ends? The first question is really the same as asking for the number of different self-avoiding walks that are possible for a given number of steps. That's easy to answer in two dimensions for an ordinary random walk: There are four choices for each step, leading to four one-step walks, 4×4 (or 16) two-step walks, and, in general, 4^n n-step walks. Similar formulas can be worked out for other dimensions.

The calculations are a little trickier for a self-avoiding random walk. In two dimensions, there are four one-step walks, 4×3 (or 12) two-step walks, and $4 \times 3 \times 3$ (or 36) three-step walks. It turns out that there are a hundred four-step walks. Calculating the number of possible five-step walks is considerably more difficult, and even with computers to help out, no one has ever gone beyond thirty-nine steps, which has 1.13×10^{17} possibilities. Many other problems involving self-avoiding walks—including determination, in different dimensions, of the typical end-to-end distance after a certain number of steps—have turned out to be difficult to solve.

"The Brownian frontier and many other examples of random motions and their intersection properties continue to be an active area of research," says the mathematician Gordon Slade of McMaster University in Hamilton, Ontario, who has worked on self-avoiding random walks as models for polymers. "Many of the remaining problems are appealing not just because of their relevance to applied fields beyond mathematics but also because the simplicity of their statements has an attraction of its own. This has drawn investigators from diverse backgrounds to study these problems, and there is hope that the progress of the recent past will continue in the coming years."

Money Matters

Curiously, the first person to discover the mathematical connection between random walks, Brownian motion, and diffusion was not Norbert Wiener but the mathematician Louis Bachelier (1870–1946)—a name not widely known in the realm of physical science. Bachelier's studies, published in 1900 in a doctoral thesis, had nothing to do with the erratic movements of particles suspended in water. Instead, he focused on the apparently random fluctuations of prices of stocks and bonds on the Paris exchange.

Because of its context, perhaps it's not surprising that physicists and mathematicians ignored or didn't even notice Bachelier's work. Those who did notice tended to dismiss his "theory of speculation" as unimportant. The mathematician Henri Poincaré, for one, in reviewing Bachelier's thesis, observed that "M. Bachelier has evidenced an original and precise mind [but] the subject is somewhat remote from those our other candidates are in the habit of treating." Instead of receiving the highest award, *mention très honorable*, which would have assured Bachelier a job in the academic community, he merited a mere *mention honorable* for his thesis.

Nonetheless, Bachelier anticipated many of the mathematical discoveries later made by Wiener and others, and he correctly foretold the importance of such ideas in today's financial markets. Bachelier himself believed in the importance of his theory, insisting that "it is evident that the present theory solves the majority of problems in the study of speculation by the calculus of probability."

In examining the role that Brownian motion may play in finance, one place to start is with the link between one-dimensional random walks and gambling. Imagine a walker starting at position zero at time zero, with coin flips deciding the direction of each step to the left or right. If the track has barriers at its two ends, which swallow up the walker to end the walk, this random process serves as an illuminating model of a famous betting dilemma called the gambler's ruin. Suppose that Ernie starts with $8 and Bert with $11. The two players repeatedly flip a coin. For each head, Bert gives Ernie $1 and for each tail Ernie gives Bert $1. The game ends when either player runs out of money.

Working out the probability that Ernie will win and the probability that Bert will win is equivalent to determining the probability of a walk ending at either barrier of a track that extends eight units in one direction from the origin and eleven units in the other. In this case, the probabilities of winning are simply 8/19 for Ernie and 11/19 for Bert (the original capital of each player divided by the total number of dollars held by both players).

Now, what if one barrier is removed and the track in that direction goes to infinity? If the walk continues long enough, it is certain to end at the remaining barrier. Thus, in betting, if Ernie plays against an opponent with an unlimited supply of capital, he will eventually be ruined. This is certainly bad news for the compulsive gambler, who, even at fair odds, faces an opponent—the entire gambling world—with virtually unlimited funds.

It's possible to plot a player's cumulative total, showing the wins and losses as a line that fluctuates up and down as the number of coin tosses increases (see also chapter 1). For increasingly long sequences of data, the line looks more and more like one-dimensional Brownian motion. To Bachelier, it also resembled the day-to-day variations in bond prices, which suggests that the market has a strongly random character. Probability, he concluded, could diffuse in the same manner as heat.

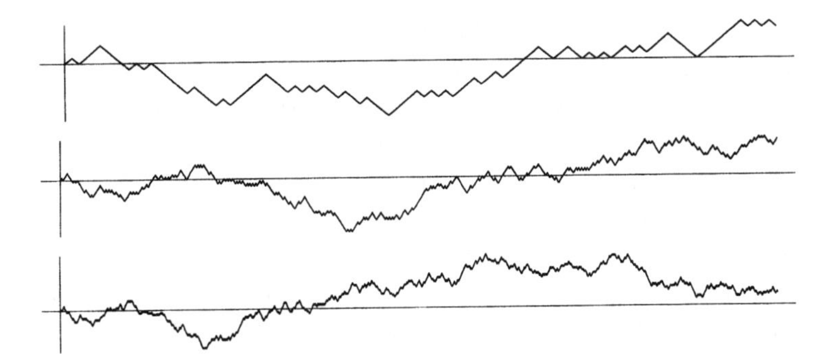

The cumulative results of honest coin tosses serve as a model of a one-dimensional random walk. Heads represents a gain of 1 and tails a loss of 1. Plotting the running total of a sequence of tosses against the number of tosses generates a fluctuating curve. As the number of tosses increases, the plots become increasingly erratic and the line appears to jiggle, much like the track of a tiny particle suspended in water undergoing Brownian motion.

In essence, Bachelier argued that charting a stock's past performance as a means of predicting its future price has little value because successive price changes are statistically independent. His approach reflected specific ideas about stock markets and their behavior, namely that the price of a stock reflects all the information available on that stock.

In this model, any variations in price would correspond to the arrival of new information. Some of that information is predictable, and some isn't. If the market works effectively, predictable information is already accounted for in the price. In other words, a rise or fall has already taken place on the basis of predictions. When the news finally arrives, it can no longer affect the price. Hence, price fluctuations occur only in response to truly new pieces of information—the parts that aren't predictable based on available data. It makes sense, then, to model fluctuations in stock prices as a Markov process, which proceeds in independent increments.

This picture of a stock market, however, isn't completely realistic, and Brownian motion generally fails as a full measure of stock-price fluctuations. Human ingenuity, fear, intuition, fallibility, and incompetence conspire to complicate the situation and put it outside the realm of pure randomness.

Interestingly, random-walk and Brownian-motion models have made a comeback in recent decades in finance theory, specifically in methods of evaluating stock options. An option is a contract by which a seller commits himself to selling specified shares within a certain time period at a price determined today. The buyer has the right not to exercise the option if this price turns out to be higher than the price at the date the option falls due. Such contracts allow holders of a large number of shares to hedge their bets. They protect themselves against a fall in price by buying the corresponding options with the idea of not using them if prices rise or remain stable. The person who sells the option takes a risk and must be compensated.

A technique for determining a fair price for such options was initially established in 1970 by the mathematician Fischer Black (1938–1995) and the economist Myron Scholes, who worked out a formula that takes into account a stock's tendency to oscillate in price. According to their model, it's not necessary to know whether a stock will go up or down in price: The direction of the price change doesn't matter. Instead, the only thing that matters is how much the stock price is likely to vary during the period of the option. A company whose stock price fluctuates a great deal over a wide range presents a bigger investment risk than one whose stock price is expected to remain relatively stable.

The Black-Scholes formula was so innovative when it was first proposed that the authors had difficulty getting their paper published. Eventually they prevailed, and soon after its publication in 1973 traders were successfully applying their pricing formula to real markets. Nonetheless, the formula represented an idealization of the behavior of markets, and researchers have since considerably refined and extended the basic model.

Finance theory has now become one of the most active fields of research in modern applied mathematics. At the heart of these developments is the constant need to evaluate assets by determining fair, rational market prices and predicting how those prices will change in consort with the prices of other goods and financial instruments. As a result, many of the mathematical tools and computational techniques already familiar to mathematicians, scientists, and engineers are finding new homes in the world of finance.

One concept that may prove of great value is the notion of scaling—the way phenomena look on different scales. The motion of a simple pendulum, for example, has an intrinsic scale because the pendulum oscillates at a particular, fixed frequency. Avalanches, on the other hand, appear to be totally random events, and they have no intrinsic scale. They come in all sizes, from just a few pebbles rolling down a hill to massive rock slides roaring and crashing down a mountainside.

In the case of avalanches, earthquakes, and even stock-market crashes, it's possible to show that the number of events goes down as the size of the events increases. Big events occur less frequently than small events, a relationship that can in many cases be expressed by a simple mathematical formula. Discovering such relationships in experimental data provides potentially valuable information about apparently random events. Similarly, fractals, which look the same on all scales, also make an appearance as geometric models of phenomena that occur on a wide range of scales. "We don't predict when an event like an earthquake will happen because it is random; all we can predict is the probability that it will happen," says the physicist H. Eugene Stanley of Boston University, a pioneer in this field. "There are a zillion problems like this, ranging from the stock market to lots of other things that are scale-free phenomena."

Galaxies and Coffee Cups

One of the most striking features of the universe, as revealed by modern telescopes, is the clustering that's apparent in the sky. Made up of more than a billion stars, our galaxy, the Milky Way, is a member of a cluster of galaxies, and this local cluster, in turn, is a member of a larger aggregation known as a supercluster. On even larger scales, clusters of clusters of galaxies appear to group into great walls, strings, sheets, and other structures, with unexpectedly large voids completely free of visible matter between these assemblages.

Interestingly, perfectly simple statistical rules can generate random collections of points that show the same self-similar clustering behavior and exhibit the same large voids evident in the distribution of galaxies in the universe. One such approach involves random walks in which the size of the steps are not fixed but vary in particular ways.

In the early part of this century, the French mathematician Paul Lévy (1886–1971) explored the possibilities and discovered a class of random walks in which the steps vary in size, from infinitesimally small to infinitely large, so no average or characteristic length can be calculated. The movements are different from Brownian motion in that a Lévy walker takes steps of different lengths, with longer steps occurring proportionally less often than shorter steps. A jump ten times longer than another, for example, would happen only one-tenth as often. It makes sense to call such excursions "flights."

In two dimensions, Lévy flights correspond roughly to a sequence of long jumps separated by what looks like periods of shorter ventures in different directions. Each "stopover," however, is itself made up of extended flights separated by clusters of short flights, and so on. Magnifying any of the clusters or subclusters reveals a pattern that closely resembles the original large-scale pattern, which means that Lévy flights have a fractal geometry—the parts on all scales closely resemble the whole (see color plate 7).

In two dimensions, the most striking visual difference between Brownian random walks and Lévy flights is the area they cover in a given time. Lévy flights sample a much larger territory than the corresponding Brownian random walks, as the following figure shows.

A similar structure in games of chance governs the pattern of successive ruins, when the player loses everything yet is given the opportunity to continue (on credit) and lose again. The resulting Lévy distribution of the frequency of successive ruins is very different from the

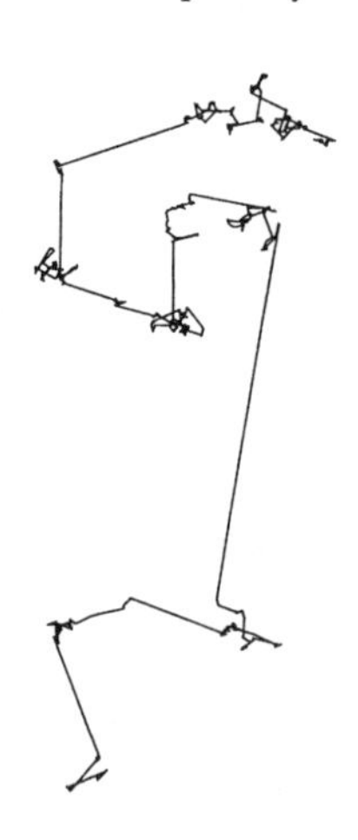

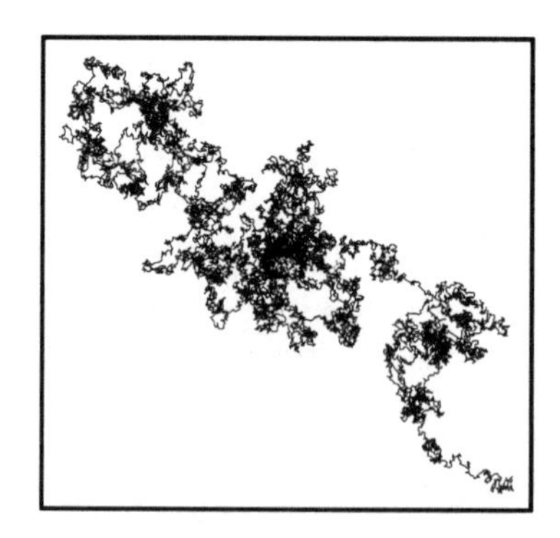

A comparison of a Lévy flight (left) with a Brownian random walk (the smudge on the top right) for the same number of steps reveals that a Lévy flight covers a considerably larger area. An enlargement of the Brownian random walk (bottom right) shows that this type of random motion lacks the long segments characteristic of a Lévy flight. (G. M. Viswanathan, Boston University)

bell-shaped symmetry of a Gaussian distribution, which characterizes ordinary random walks. In the Lévy case, the distribution is strongly skewed, with a very long tail that drops off to zero very slowly as the size of the events increases.

The mathematician Benoit B. Mandelbrot of IBM and Yale University originally learned about these different random walks from Lévy himself, and Mandelbrot later extended and applied Lévy's ideas in his formulation of fractal geometry. Mandelbrot found that he could use Lévy flights to create convincing portraits of the distribution of visible matter in the universe. He simply erased the long jumps and made each stopover represent a star, a galaxy, or some other blob of matter. The resulting pattern of clustered spots, each of which in turn is made up of subclusters, resembles the sheets, bubbles, and other aggregations of galaxies evident in astronomical observations. Of course, Mandelbrot's model doesn't necessarily account for the way galaxies actually formed in the universe, but it does suggest the kind of structure that may be present.

Lévy flights and the statistics associated with them also provide useful models of turbulent diffusion. If you add a drop of cream to your coffee without unduly disturbing the liquid, the random motion of the molecules slowly spreads the cream into the coffee. Stirring, however, adds turbulence, and the liquids mix much more rapidly. Mathematically, it's possible to think of turbulence as the combined effect of a large number of vortices—whirlpools of all sizes and strengths. Any particles (or molecules of the constituents of cream) caught in such whirlpools would be rapidly separated from one another and dispersed. A plot of changes in the distance between two initially adjacent particles would look much more like a Lévy flight than a Brownian random walk.

Lévy flights can arise out of chaotic systems, in which a sensitive dependence on initial conditions plays a crucial role (see chapter 7), and out of random systems, reflecting the same sort of haphazardness shown in Brownian motion. New statistics based on Lévy flights must be used to characterize these unpredictable phenomena. Such models may be useful for describing, for example, the transport of pollutants and the mixing of gases in Earth's atmosphere. "In these complex systems, Lévy flights seem to be as prevalent as diffusion is in simpler systems," notes the physicist Michael Shlesinger of the Office of Naval Research, who pioneered the application of Lévy statistics to turbulent diffusion and other physical phenomena.

In both the atmosphere and the ocean, fractal patterns associated with turbulence may have a strong influence on ecosystems, affecting, for example, the foraging behavior of certain birds. Weather systems and the distribution of plankton, krill, and other organisms in the ocean may guide the flight patterns of the wandering albatross.

Flights of the Albatross

The wandering albatross (*Diomedea exulans*) flies extraordinary distances in search of food. Riding the wind on long, thin, rigidly outstretched wings, it skims the waves as it glides for hours over the ocean surface. Truly a world traveler, this seabird regularly circles the globe at southern latitudes, plunging into the sea to scoop up squid and fish along the way and sometimes following cruise ships and other vessels to pick up scraps thrown overboard. Its white plumage, white beak, black wing tips, and wingspan of eleven feet or more make it a dramatic sight in the sky. In one week, it can travel thousands of kilometers on a single foray to gather food for its baby chick.

Biologists at the British Antarctic Survey in Cambridge, England, have been investigating the role of seabirds and seals as the top predators in the marine food web of the southern ocean. Their long-term goal is to assess the impact of these animals on the ecosystem. As one component of this effort, the biologist Peter Prince and his coworkers at the British Antarctic Survey equipped wandering albatrosses with electronic activity recorders or radio transmitters for satellite tracking to determine the birds' foraging behavior and identify patterns in the way they search for food.

In one experiment, the researchers attached electronic activity recorders to the legs of five adult birds, who made nineteen separate foraging trips. The devices recorded the number of fifteen-second intervals in each hour during which the animal was wet for nine seconds or longer. The wet periods indicated interruptions in a bird's flight path when it alighted on the water to eat or rest.

To cope with the large quantity of data generated by such studies, the biologists enlisted the help of a team of physicists to identify patterns in the way albatrosses search for food. They turned to H. Eugene Stanley and his colleagues Gandhi Viswanathan and S. V. Buldyrev at Boston University.

According to the analysis performed by the Boston group, the data show that the flight patterns of wandering albatrosses—as they fly, settle on the sea, then fly off again—fit the type of random motion characteristic of a Lévy flight, in which the birds make long journeys interspersed with short foraging flights clustered in a small area. Ecologists speculate that the flight patterns of the wandering albatross have evolved to exploit the patchy distribution of fish and squid, which may reflect the distribution of plankton in the restless ocean, and this patchiness in turn may arise from ocean turbulence effects.

Such patterns may also occur in other biological systems. Some scientists have recently applied Lévy random walks and Lévy statistics to the foraging behavior of ants and bees. Others are studying possible uses of these models in physiology and medicine, including the characterization of heartbeat rhythms and the branched structure of the lung's airways.

It's possible that in biological systems there's an evolutionary advantage to having Lévy statistics. Because the environment appears to be fractal, an organism that behaves fractally can better take advantage of such patchy opportunities. For the wandering albatross, this means wide-ranging, stop-and-go searches for food that may be unpredictably scattered across the ocean.

"Newtonian physics began with an attempt to make precise predictions about natural phenomena, predictions that could be accurately checked by observation and experiment," Michael Shlesinger and two colleagues note in a 1996 essay entitled "Beyond Brownian Motion." They further explain, "The goal was to understand nature as a deterministic, 'clockwork' universe." The application of probability to physics developed much more slowly, they contend. The earliest uses of probability were in characterizing data—how the average represented the most probable value in a series of repeated measurements, and how the various measured values fitted the familiar bell-shaped curve of a Gaussian distribution.

In the nineteenth century, such distributions also came to represent the behavior of molecules in a gas. In constant motion, molecules repeatedly collide and change speeds. At a given temperature at any moment, however, the largest number of molecules travel at a well-defined characteristic speed, and the distribution of speeds among all the molecules fits a specific type of curve.

Thinking about random walks and Brownian motion, about Lévy flights and chaos, and about Markov processes and fractals means tangling with the complex interplay of pattern and randomness in nature and in human activity. It's no simple matter to analyze phenomena that change randomly and uncontrollably from moment to moment, whether these changes are physical, biological, or mathematical. Making such efforts, however, begins to cast light on the essence of life, which seems to teeter far out of equilibrium and perpetually on the brink of randomness.

9

Gambling with Numbers

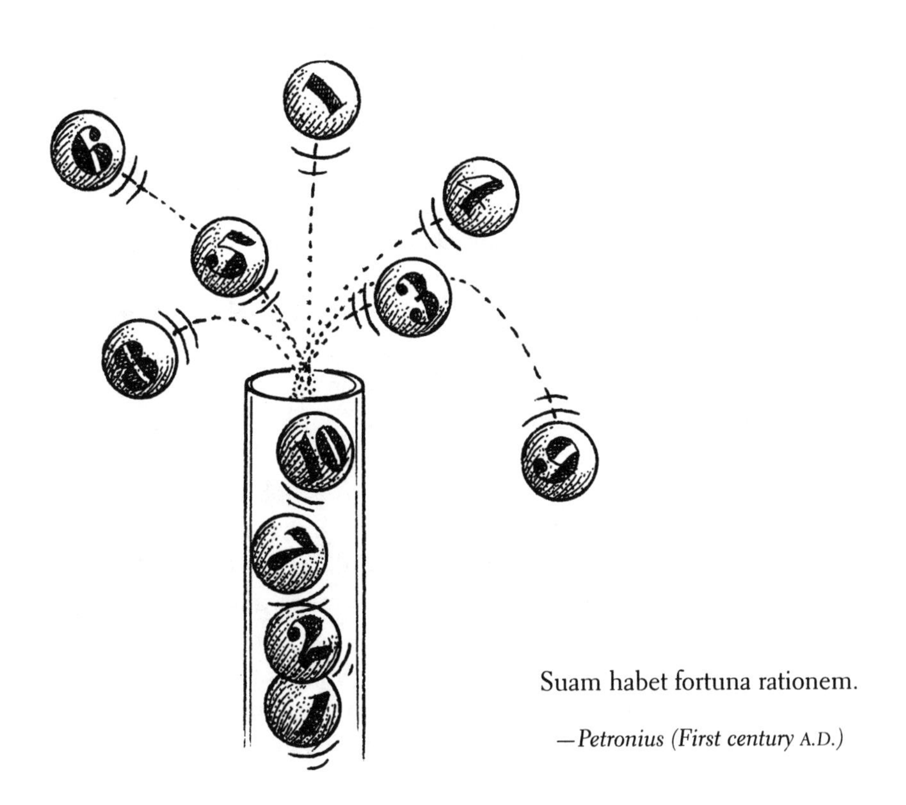

Suam habet fortuna rationem.

—*Petronius (First century* A.D.*)*

Three coins clink into a slot. A gnarled finger stabs the spin button, setting three reels in rapid motion. The bleary-eyed player stares intently as the whirling reels finally begin to slow down, then click into position. The display shows a row of two sevens and a bar. No jackpot this time.

Farther down the bank of slot machines, another gambler screams with delight as a shower of coins erupts from a chute. Bells clang and lights blink furiously to announce the lucky strike. Overhead, the jackpot total emblazoned on an electronic display compounds by the second, numbers changing as quickly as the digits of a zooming rocket's odometer.

Slot machines have been a mainstay of casinos and other gambling venues for more than a century. Until the last few decades, these machines were purely mechanical, relying on an array of gears, springs, levers, and weights to govern the motion of the three spinning reels. Typically, each reel displayed a ribbon of symbols, ranging from various types of fruit to bars and numbers. It could stop at any of twenty positions, giving a total of eight thousand different combinations for the three reels together.

The mechanism's precarious balance of gears and weights provided the sensitive dependence on initial conditions necessary for its random selection of stopping positions. However, players sometimes managed to manipulate or take advantage of a malfunctioning or ill-adjusted machine. By pulling the handle in just the right way or inserting the coin at the right moment, they could influence the reels' motion and increase the likelihood of success.

Nowadays a slot machine's interior looks a lot like the inside of a desktop computer. A mixture of software and instructions encoded into chips controls the machine's actions. Circuitry links the electronics to digital displays, mechanisms for paying out coins, and small motors to drive the reels.

The main role of the machine's microprocessor is to generate random numbers, which can be assigned to any combination of symbols on the reels. In effect, the microprocessor dictates what the machine displays, how much it pays out, and how often. This repeated selection of random numbers continues even when the machine isn't being played. By pushing the spin button or pulling the handle, a player merely activates the display. The reels spin, but their eventual stopping positions have already been determined by the microprocessor. Indeed, the reels of a modern slot machine serve no purpose other than to keep the player entertained.

The use of microprocessors allows slot-machine operators to manipulate the odds. In a typical mechanical model, each of the eight thousand possible combinations has the same chance of coming up, so the maximum odds of a payoff are eight thousand to one. In contrast, the programmable electronic version can offer much higher odds (and correspondingly bigger payoffs). One simply assigns fewer numbers to combinations that one wants to come up less often. For the top jackpot, the odds of pulling the single right combination may be as high as ten million to one.

In fact, slot-machine operators often go a step further by networking hundreds of machines and using telephone lines to coordinate the odds and payoffs. Each time a machine is played, the potential winnings of the combined jackpot rise. Such schemes have yielded payoffs of more than $11 million.

To simulate chance occurrences, a microprocessor or computer can't literally toss a coin, roll a die, or spin a roulette wheel. The behind-the-scenes agent responsible for the illusion of chance is a special numerical recipe built into the software to generate apparently patternless strings of digits. Such formulas produce random numbers not only for slot machines but also for video poker and other casino games. Moreover, electronically manufactured random numbers play a key role in any computer game that involves chance and, for many scientists, in computer simulations of physical processes.

The trouble is, just as no real die, coin, or roulette wheel is ever likely to be perfectly fair (see chapter 1), no numerical recipe produces truly random numbers. The mere existence of a formula suggests some sort of predictability or pattern. After all, the computer simply follows a set procedure to generate the numbers, and restarting the process typically produces exactly the same sequence of digits. Moreover, the sequences eventually begin repeating themselves. In principle, every result could be predicted in advance. "Chance has its reason," the Roman satirist Gaius Petronius once remarked in a somewhat different context.

By the Cards

The playing cards slip instantly, precisely, and soundlessly into place. No smudges or creases ever mar the crisp, bright faces heading the unnaturally tidy columns and rows. With no deck of cards to handle, shuffle, deal, sort, stack, or position, there's only the point and click of a mouse pushed across a pad beside the computer's keyboard. This is solitaire on the computer screen. All the chance involved in the game is bound up with how the software selects random numbers and translates those numbers into specific arrangements of cards. Indeed, nearly every type of computer and many computer programs have a built-in set of instructions for producing random numbers.

Scientists, engineers, and others use random numbers for tackling a wide range of problems, from modeling molecular behavior and sampling opinion to solving certain equations and testing computer programs. Such numbers also play a crucial role in computer graphics for rendering realistic-looking images, such as mottled surfaces and other textures.

A century ago, researchers who needed random numbers in their scientific work actually tossed coins, rolled dice, dealt cards, or picked numbered balls out of urns. They also combed census records, directories, and other lists to obtain the digits they required. In 1927, L. H. C. Tippett, a member of the Biometric Laboratory at University College in London, went to the trouble of compiling and publishing a table of 41,600 numbers, obtained by taking the middle digits of the measured areas of parishes in England, to serve as a convenient source of random numbers.

Nearly thirty years later, experts at the RAND Corporation used a special machine—in effect, an electric roulette wheel—to generate random bits (1s and 0s), which were then converted into a table of a million random decimal digits and published as a book. To remove slight biases uncovered in the course of rigorous, intensive testing and attributed to faulty wiring, the digits were further randomized by adding all pairs and retaining only the final digit.

The introduction of computers in the late 1940s spurred the development of mathematical procedures, or algorithms, that employ random numbers. One of the first applications of this method was computing the paths of neutrons in a thermonuclear explosion, in which the fission of a uranium jacket triggers nuclear fusion in a hydrogen core. The idea was to follow the progress of a large number of neutrons as they bounce around, get absorbed, trigger nuclear fission, or escape from the bomb material.

Each stage of the computation involved a decision based on statistical probabilities derived from the physics and geometry of the situation. A given neutron started with a randomly selected velocity and position. When it collided with something, random numbers decided the outcome. For example, in a simple collision, in which the neutron merely rebounds, random numbers would determine the particle's new speed and direction. If a fission occurred because the neutron happened to be absorbed by a nucleus, making it unstable, a random-number distribution appropriate to the relevant physics would specify the number of neutrons freed by the interaction. These neutrons, in turn, would each be followed in the same fashion as the initial neutron.

In effect, the computer played a game of chance—a kind of nuclear pinball—and the neutrons took random walks (see chapter 8). Repeated for a sufficiently large number of neutrons, the computations would add up to a statistically valid picture of the conditions necessary for nuclear fission.

Stanislaw Ulam (1909–1984), a mathematician at Los Alamos, New Mexico, was one of the first to realize that computers would allow such computations on a sufficiently large scale to be useful to scientists and engineers designing and building nuclear bombs. Ulam himself had long demonstrated a keen interest in random processes. He found it amusing to cite the number of times he had driven into a filled parking lot just as someone was leaving a space. He played solitaire to relax, and he thoroughly enjoyed poker. In fact, the use of statistical sampling to calculate the probability of winning a game of solitaire is what led Ulam to suggest using a similar strategy in physics problems.

The outcome of a game of solitaire depends upon the initial distribution of cards in the deck. The number of different ways of arranging fifty-two cards is $52 \times 51 \times 50 \times 49 \times \ldots \times 3 \times 2 \times 1$ (often written as 52!, where the exclamation mark means "factorial"). The total number of combinations is a number seventy digits long! Instead of checking all the possibilities and counting up the winning distributions, Ulam's strategy was to examine the outcome of just a few thousand shuffles of the deck. He realized that a computer can easily be programmed to shuffle a deck, with cards represented by the numbers 1 to 52, and to play a game of solitaire repeatedly with a randomly chosen assortment of cards. From the number of successes in the various trials, he could then estimate the probability of winning.

In 1947, Ulam and John von Neumann (1903–1957), a brilliant mathematician at the Institute for Advanced Study in Princeton, New Jersey, proposed that the Los Alamos researchers use computers to try a statistical approach in physics calculations. Since Ulam had an uncle who would often borrow money from relatives because he "just had to go to Monte Carlo" to visit the Mediterranean principality's famed roulette wheels, Ulam and von Neumann chose to call their technique the Monte Carlo method. The name has endured as the term applied to any mathematical procedure or algorithm that employs statistical sampling and random numbers.

To take advantage of the method, however, researchers needed a quick, efficient way to obtain vast stocks of random numbers for use in their computer programs. Von Neumann was one of the early leaders in developing computer-based random-number generators that transform the digits or bits of a given number in such a way that the new number appears to be independent of previously generated numbers.

In von Neumann's "middle-square" algorithm, the idea is to start with an arbitrary n-digit number, or "seed." Squaring the number yields a new number with roughly twice as many digits. The middle n digits of the first answer become the first random number. Those digits in turn serve as the seed for generating the second random number, and so on. The result is a string of whole numbers that looks like a random sequence. For example, squaring the initial seed 6744 gives 45481536. The middle digits 4815 become the next seed value. Continuing the process produces the following sequence of four-digit numbers: 1842, 3929, 4370, 0969, and so on. The sequence looks patternless. After a few more repetitions, however, the procedure generates the value 2500. When 2500 is squared, the result is 06250000, yielding 2500 again! The sequence gets stuck on 2500.

It turns out that many other seeds also lead to sequences that eventually get trapped at a single value or in a short repeating cycle. For instance, if 2100 comes up, the values cycle through 4100, 8100, 6100, and back to 2100 from then on. Moreover, with just four digits, one can get no more than ten thousand different four-digit numbers, and that's clearly not enough for many purposes. One would also like to obtain a uniform distribution, in which all the numbers from 0000 to 9999 appear equally often. In the long run, however, the numbers 2100, 4100, 6100, and 8100 come up far more frequently than any others.

Increasing the number of digits doesn't overcome such fundamental flaws, so Neumann's middle-square method has largely been abandoned. Mathematicians and computer scientists have since developed more sophisticated recipes for generating sequences of random numbers, but they remain stuck with the disquieting notion that their methods are not foolproof—that patterns can appear unexpectedly among a sequence's members. "There is no such thing as a random number—

there are only methods to produce random numbers, and a strict arithmetic procedure, of course, is not such a method," von Neumann declared about forty years ago. He was right, but that didn't keep mathematicians and others from trying to do better.

If a sequence of numbers were truly random, then knowing any one number in the sequence would give you no information that could help you work out any other number. That rule clearly isn't obeyed in a method involving a procedure in which the result of one step is the seed for the next. Knowing one value allows you to predict every subsequent value. Hence, the numbers generated by computer-based methods are generally referred to as pseudorandom, and researchers have focused on searching for the best possible recipes to use with particular applications, knowing full well that a given scheme may occasionally fail to produce the requisite amount of randomness.

The renowned computer scientist Donald Knuth of Stanford University tells the story that he once tried to create "a fantastically good" random-number generator and invented a particularly intricate thirteen-step method to do the job. His intent was to make his algorithm so complicated that someone reading a listing of the computer program would not know what it was doing without explanatory comments.

It seemed plausible to Knuth that his complicated method, with its digital twists and turns, would produce "at least an infinite supply of unbelievably random numbers." It didn't. On the very first try, the sequence of generated numbers converged to the ten-digit value 6065038420 and stayed there. When he tried another starting number, the sequence began to repeat after 7401 values. "The moral of this story is that random numbers should not be generated with a method chosen at random," Knuth cautioned. "Some theory should be used."

Scrambled Bits

When the reels of a slot machine finally grind to a halt to display a row of cherries, bars, or sevens, the random-number generator operating behind the scenes to decide the outcome is usually of a type invented in 1951 by the mathematician Derrick H. Lehmer (1905–1991). Known as "linear congruential" random-number generators, these procedures are among the most commonly used methods for scrambling digits.

Typically, a starting number of a given length (usually limited by the size of numbers a computer can handle) is multiplied by some constant. Another constant is added to the result, and the answer is then divided by a third constant. The remainder after division constitutes the required pseudorandom number, and it also becomes the seed for the next round of steps (see appendix for examples).

With a proper choice of multiplier and divisor, such a generator produces a sequence of numbers that are difficult to distinguish from random numbers generated purely by chance, even though a sequence can display striking regularities under certain conditions. Despite such flaws, one widely used variant is random enough to run the state lotteries and the casinos in Las Vegas and Atlantic City with no one the wiser. Because congruential generators can produce numbers no bigger than the selected divisor, however, their output is limited to sequences of no more than a few billion or so random numbers. That's not enough for many research applications.

As scientists turn to increasingly speedy and powerful computers, they run the risk of using up all the pseudorandom numbers that a computer's standard generator can produce. On a desktop computer, it's possible to exhaust the generator with a run that lasts only a few days. Furthermore, if too many random numbers are drawn from a relatively small pool, the selected numbers as a group may no longer pass some of the more stringent tests of randomness.

In 1991, George Marsaglia and Arif Zaman of Florida State University introduced a new class of random-number generators that produce astonishingly long sequences of pseudorandom numbers before the numbers start repeating themselves. Indeed, the Marsaglia-Zaman generators act like randomness amplifiers. Some have periods of at least 10^{250} before sequences begin to repeat themselves (see the appendix on page 205 for examples).

A long period by itself isn't enough, however. The new generators also proved to be remarkably efficient and didn't take up inordinately large amounts of a computer's memory. In addition, they passed a battery of stringent randomness tests. So, the Marsaglia-Zaman random-number generators quickly attracted the attention of scientists who were interested in high-quality, long-period sources of random numbers. Even gambling experts came calling.

However, high-quality, well-tested random-number generators can still yield incorrect results under certain circumstances, as one group of researchers discovered, to their surprise and chagrin, when they used one of the new Marsaglia-Zaman procedures in a Monte Carlo study of physics related to magnetism.

Flawed Flips

From the ceramic butterflies used for mounting children's drawings on refrigerator doors to the iron cores of audio speakers and telephone receivers, magnets are a significant component of everyday life. Their striking behavior has also been a source of endless fascination for millennia.

If you cut an iron bar magnet in half, you end up with two smaller magnets. Like the original magnet, both pieces have north and south poles. When the two magnets are brought together, opposite poles attract each other and like poles repel. Cutting each of the magnets in half again gives a total of four magnets, and so on, down to the individual atoms of iron.

All magnets have a magnetic field—often depicted as lines representing the ability of a magnet to exert a force and interact with other magnets. The fields, in turn, are generated by the motion of electrons. In an electromagnet, the electrons typically stream along a copper wire. In a so-called permanent magnet, they are tethered to atoms, spinning as they orbit their nuclei.

The orbits and spins of electrons in a metal determine whether the metal can become a permanent magnet. Like whirling tops, electrons can spin either clockwise or counterclockwise, producing magnetic fields that point up or down. In many atoms, electrons occur in pairs of opposite spin, which makes their magnetic fields cancel out. In contrast, atoms of iron, cobalt, and nickel contain lone, unpaired electrons whose magnetic fields are not canceled out, so each atom acts like a tiny bar magnet with its own north and south poles.

In 1907, the French physicist Pierre Weiss (1865–1940) proposed that a magnetizable material consists of many small regions called domains. Atoms within a domain are lined up so that all of their magnetic fields point in the same direction. Each domain is, in effect, a microscopic magnet.

In a piece of unmagnetized iron, the magnetic fields of the domains point in different directions. Magnetizing iron by exposing it to a powerful magnetic field forces atoms in the individual domains to line up in one direction. The combined effect of the newly aligned domains gives the chunk of iron its own magnetic field. Heat can demagnetize a magnet by jumbling the domains and destroying the regimentation responsible for the overall magnetic effect.

Real magnets are complicated materials. So, to understand how magnetization occurs, physicists have over the years developed mathematical models that capture key features of the interactions among atoms and their spinning electrons. The so-called Ising model represents a magnetic system stripped to its barest bones—nothing more than an orderly array of spins (usually drawn as little arrows pointing up or down). In one dimension, the spins are evenly spaced along a line; in two dimensions, at the corners of the squares of a checkerboard grid; and in three dimensions at the vertices of a cubic lattice. In an additional simplification, only neighboring spins are allowed to influence each other.

Though the model is unrealistic, it has the advantage that scientists can express it in mathematical form and try to solve the resulting equations to predict how arrays of spins behave at different temperatures. The first researcher to do so was Ernst Ising, who worked out the exact results for spins lined up in a row. In 1944, Lars Onsager (1903–1976) solved the problem in two dimensions and demonstrated that below a certain temperature spins spontaneously line up in one direction, going from a disordered to an ordered state.

Despite its simplicity, the Ising model reflects some important properties of real magnets. At high temperatures, spins fluctuate rapidly and the material displays little magnetism. As the temperature falls, the spins become organized into domains that share the same spin, and the domains grow larger as the system cools further. (This is illustrated in the following figure.)

No one has been able to find an exact solution to the equations of the three-dimensional Ising model, so researchers have turned to computer simulations and various Monte Carlo methods to get approximate answers. They start with a mixture of up and down spins locked into a lattice. Then they select a site at random and compare its spin with the spins of its immediate neighbors. If necessary, the spin is reversed (flipped from up to down or down to up) to bring it into alignment with the majority of its nearest companions. A typical simulation can require millions or even billions of such steps.

Lowering the system's temperature decreases the probability of a spin flip, so it takes longer for an array of spins to reach a state of equilibrium. Moreover, as the temperature gets closer to that at which there is a transition to a state in which all the spins are in the same direction, the spin fluctuations get slower and slower—as if the system were congealing—and it takes more and more calculations to determine the eventual outcome. Even the best of today's supercomputers are too slow to calculate the spin-by-spin details of the transition.

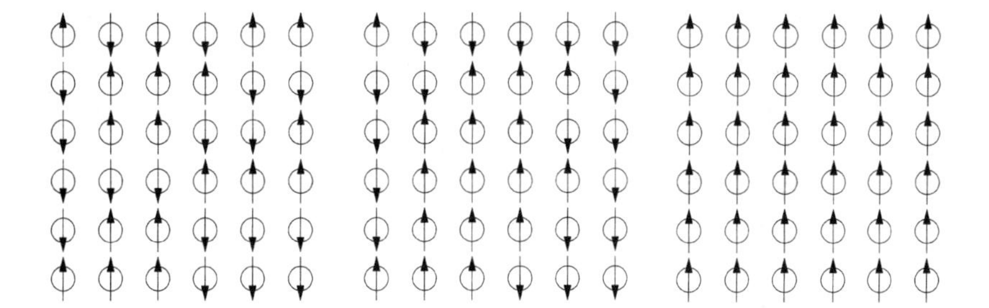

As a simple model of a magnetic material such as iron, imagine an array of particles, each one spinning either clockwise or counterclockwise so that its spin axis (or north pole) points up or down. In the Ising model, particles can interact with their immediate neighbors, changing their spin directions to match those of their neighbors with a probability that depends on the temperature. At high temperatures, the spins are randomly oriented up or down (left). As the temperature decreases, neighboring spins tend to become aligned (middle). In magnetized iron, the spins are all lined up in the same direction (right).

A few years ago, Ulli Wolff of the University of Kiel in Germany introduced an algorithm that circumvents the computational slowdown. Instead of flipping spins individually, his Monte Carlo method manipulates clusters of spins, which saves a considerable amount of time on the computer.

In 1992, Alan M. Ferrenberg, a computational physicist at the University of Georgia, and his coworkers David P. Landau and Y. Joanna Wong began testing the Wolff spin-flipping algorithm using one of the new Marsaglia-Zaman generators as a source of random numbers. In preparation for simulating the three-dimensional Ising model, Ferrenberg tested the program on the two-dimensional version, which has a known answer. He got the wrong result.

Initially convinced that the problem lay in how he had written his computer program, Ferrenberg spent three weeks looking for errors but found none that affected the result in the way he had observed. As a last resort, he tried different random-number generators and, to his surprise, found that he came much closer to the correct answer by using a linear congruential generator, which has known defects.

For Ferrenberg, this outcome was both discouraging and disquieting, and the experience served as a stern warning. It's difficult to foretell when a particular simulation or calculation may unexpectedly trip over some peculiarity of a given random-number generator, no matter how highly regarded its pedigree is.

Identifying such situations gets particularly difficult when researchers venture into new terrain where the answer isn't known beforehand. All one can do is repeat the calculations with different random-number generators. Deciding which one has given the "correct" answer, however, can be problematic. The real trouble, however, is that many researchers fail to take even this basic precaution and end up building their conclusions on potentially shaky foundations.

To Marsaglia and Zaman, the experience of Ferrenberg and his colleagues represents a pathological case. Most of the time, conventional methods work remarkably well, whether for playing the game of Donkey Kong or for devising sophisticated simulations in high-energy physics. "Standard random-number generators have been available in most computer systems for some forty years, and it is a rare problem for which they give poor results," they note. Nonetheless, "every deterministic sequence will have problems for which it gives bad results, and a researcher should be able to call on a variety of random-number generators for comparing results." Marsaglia and Zaman wryly conclude, "While interesting theory can lead to new kinds of random-number generation, in the end it is, alas, an empirical science."

Every scheme now in use for generating random numbers by computer has some sort of defect. By sharing experiences and comparing notes, researchers can develop guidelines for when to use and when not to use particular generators. As the mathematician Robert R. Coveyou at the Oak Ridge National Laboratory in Tennessee once said, "The generation of random numbers is too important to be left to chance." John von Neumann put it a little more strongly: "Anyone who considers arithmetical methods of producing random digits is, of course, in a state of sin." Marsaglia and Zaman add, "Only through collective experience can we hope to reach a state where we may 'go and sin no more.' "

Disks and Diehard Batteries

In the spring of 1996, on the eve of his retirement from active duty at Florida State University, George Marsaglia started giving away random numbers. In a project funded by the National Science Foundation, he had packed his hard-won expertise onto a CD-ROM to create a catalog of random bits for the information age, and the disks were finally ready.

Marsaglia's primary purpose was to provide a source of what he termed "unassailable random numbers" for serious Monte Carlo studies and simulations, where billions of such numbers are often required. In smaller doses, his apparently patternless strings could also be used to provide reliable and verifiable random choices for designing experiments and clinical trials, for running sample surveys and choosing jury panels, and in other applications of a statistical nature.

The Marsaglia Random Number CD-ROM contains some five billion random bits, divided into sixty ten-megabyte files, which can be read one, eight, sixteen, or some other number of bits at a time, depending on the application. The random bits were made by combining three sources of electronic noise with the output from the best of the latest crop of deterministic random-number generators developed by Marsaglia and Zaman.

Because the result of combining a truly random string with any other bit string—no matter how patterned—is still random, Marsaglia also mixed digital tracks from rap and classical music and even a few digitized pictures into some of the ten-megabyte files on the CD-ROM. Both the unadulterated and the mixed files passed Marsaglia's "Diehard" battery of randomness tests, which are also on the disk.

Loosely speaking, any number that belongs to a sequence of random numbers is a member of that string by chance. Ideally, it has no connection with any other number in the sequence, and each number has a specified probability of falling in a given range. If the probability distribution is uniform, one would expect that each of the ten digits from 0 to 9 occurs about one-tenth of the time. For binary digits, 1s and 0s should come up equally often in the long run.

It's also useful to check whether the pairs of digits 00, 01, 10, and 11 each come up roughly one-quarter of the time, and whether the triples 000, 001, . . . 111 each appear about one-eighth of the time. There are equivalent checks for larger groups of bits and for various groupings of decimal digits.

Such traditional tests, however, aren't sufficient to guarantee a reasonable degree of randomness. Over the years, Marsaglia has developed additional tests to weed out hidden patterns. For example, in one of his Diehard tests he considers a string of random bits as a succession of overlapping twenty-letter "words," made up of the "letters" 1 and 0. The first word starts with the first bit of the string, the second word with the second bit, and so on. Given that there are 2^{20} possible twenty-letter words, the test involves examining the first two million overlapping words derived from the random string to count how many of the possible words never appear in the sequence. For a truly random sequence, the number of missing words should be very close to 142,000, according to probability theory.

Marsaglia calls this a "monkey test," based on the proverbial monkey sitting at a typewriter and randomly hitting keys (see the preface). "It's one of my most stringent tests," he says. "No commercial electronic generator of random bits that I know of can pass this test." Interestingly, generators that rely on electronic noise to produce random digits also fail. On the other hand, the best of the new Marsaglia-Zaman generators do pass all the tests in his battery, including the various monkey tests.

The playful names of other tests hint at alternative strategies for rooting out patterns in pseudorandom sequences: the birthday-spacing test, the count-the-1s test, the parking-lot test, the minimum-distance test, the overlapping-sums test, and the craps test. Checking for randomness is difficult partly because it's the nature of randomness that anything can happen (see color plate 8). Ironically, if a generator produces sequences of pseudorandom numbers that always pass a given test, then it's probably flawed!

When Alan Ferrenberg and his colleagues discovered that one of the superior Marsaglia-Zaman random-number generators failed to give the correct answer in their simulations of the Ising model, they had also, in effect, tested the random-number generator. But they weren't sure why the generator had failed. A year or so later, a group of researchers in Finland formulated several tests that pinpointed the problem.

One of these tests involved a random walker—a point hopping from one quadrant to another of a square split into four equal sections, as dictated by a string of random numbers. If the movements were truly random, the point would spend an equal amount of time in each quadrant. The same random-number generators that failed to work for Ferrenberg also failed this simple test.

One alternative to relying on computer-based methods is to take advantage of unpredictable events in physical systems, such as the static heard on a radio between stations and during thunderstorms or the sporadic clicks of a Geiger counter monitoring the radioactive decay of atomic nuclei. Such a strategy, however, has its own difficulties—from a lack of reproducibility to potentially false assumptions about the randomness of the physical process itself.

Natural Numbers

A grizzled prospector warily treads a narrow, rugged path along the edge of a cliff. His Geiger counter registers a few sporadic clicks as he proceeds toward a rocky outcropping. A few more steps take him next to the rock, and the counter begins to clatter furiously, broadcasting the random decays of uranium and other radioactive atomic nuclei embedded in the rock.

An atom of the most common form of uranium has a nucleus that consists of 92 protons and 146 neutrons. Because the nucleus is unstable, every once in a while it spontaneously breaks up into two unequal parts—a small fragment known as an alpha particle, which is made up of two protons and two neutrons, and the remainder of the original nucleus. A Geiger counter detects the ejected alpha particles—one for each decay of a uranium nucleus.

The theory of quantum mechanics, which describes the behavior of atoms, electrons, and nuclei, posits that the instant at which an atom decays occurs by chance. There's no detectable difference between a uranium atom that is about to decay and one that's still far from splitting up. There's no warning sign or yellow light signaling the approaching event. All a physicist knows is that half of the uranium atoms present in a given sample will decay in 4.5 billion years, and then half of the remainder in another 4.5 billion years, and so on.

One way to convert the decays of a radioactive element such as uranium into a string of random numbers is to measure the time between decays as detected by a Geiger counter. One can then compare the times between successive pulses, taking two intervals at a time. If the first interval is longer than the second, the generated random bit is 0, and if the second is longer than the first, the bit is 1. Thus, the procedure can be used to simulate a coin toss or to generate a random string of bits. The method isn't foolproof, however. Although each decay in a radioactive source may occur randomly and independently of other decays, it's not necessarily true that successive counts in the detector are independent of each other. For example, the time it takes to reset the counter between clicks might depend on the previous count. Moreover, the source itself is changing in composition, as one element is transformed into another. Even the way a detector's electronics heats up during use can introduce a bias.

The crucial underlying problem is deciding whether the physical process itself is, in fact, random. In the case of physical randomizers such as coins, dice, and roulette wheels, biases of various kinds do crop up, and gamblers can take advantage of these vestiges of pattern to make informed guesses about probable outcomes.

In the late 1970s, a group of professors and graduate students at the University of California at Santa Cruz did just that when they developed an ingenious computer system to surreptitiously record and analyze data from roulette wheels at casinos in Las Vegas and obtain advice on which numbers to play. Despite a variety of technical glitches, they eventually made their scheme work and succeeded in beating the casinos.

Along the way, the researchers found that what they learned from careful timing of a ball's journey around a wheel's rim before it drops into a pocket could also be gleaned by a croupier who operates the same roulette wheel five nights a week year after year. The operator gets to know intimately the slight imperfections of the track, the characteristic bounce of different kinds of balls, the heft of the rotor, and its drag on the central spindle. After years of practice, it's not impossible that a croupier could learn to flick the ball up on the track and make it drop from its orbit into a pocket on a given half of the wheel—and get it right more often than not.

It's interesting to note that several of the participants in the Santa Cruz escapade were also important figures in bringing the notion of deterministic chaos and the idea of sensitive dependence on initial conditions (see chapter 7) into the lexicon of physical science. In fact, it's relatively easy to design and build a simple electronic device that can exhibit behavior ranging from completely predictable to chaotic (with predictability in the short term) to apparently random. The setting of a single knob decides the level of unpredictability.

For example, one can readily build a device that flashes a light at a rate determined by the voltage. Pressing a switch freezes the light either on or off. At low voltages, the flashes occur slowly enough that one can choose whether the light ends up on or off. At higher voltages, the flashes occur more frequently, and the results of pressing the switch become dependent upon one's reflexes. At even higher voltages, the result of pressing the switch appears completely unpredictable in the same way that coin flipping looks random because we don't know and can't control the initial conditions of each flip precisely enough.

It's possible to improve on this device by deriving randomness not from the flashes of a light bulb or a light-emitting diode (LED) but from the motion of electrons in electronic components. One can then amplify this electron noise and make the infinitesimal jiggling of individual electrons visible in the erratic flashing of an LED. As in the case of the radioactive decay of individual nuclei, chance rules the behavior of matter at the submicroscopic level.

Thus, physicists generally regard the emission of light by an excited atom as a random process. They assume, for example, that it's impossible to predict precisely when one of an atom's electrons will drop from a higher to a lower energy level, emitting a photon in the process. But such randomness doesn't follow automatically from any other fundamental principles of quantum theory.

To test quantum processes for randomness, Thomas Erber of the Illinois Institute of Technology has analyzed data obtained from researchers at the National Institute of Standards and Technology in Boulder, Colorado. The NIST scientists studied light absorbed and emitted by a single mercury ion held in a special trap and irradiated continuously by laser light of a certain frequency. In response, the suspended mercury ion flashed on and off at irregular intervals, acting like an optical telegraph.

Applying a variety of techniques developed in recent years to search for patterns in encrypted data and other strings of digits, Erber studied a sequence of twenty thousand numbers representing the time intervals between successive flashes in the NIST data. He detected no discernible patterns in the limited data available. So, quantum mechanics passed this particular set of randomness tests.

It's possible that one of the simplest of all physical systems—a single atom suspended in a trap and tickled by laser light—may turn out to be a source of truly random numbers.

Luckily, standard computer-based random-number generators are good enough for most practical purposes. In a modern slot machine, even a linear congruential generator will suffice to keep a player from taking advantage of any subtle patterns that may be present. There's also nothing we can do mechanically—when pulling the handle, pushing the button, or slipping in the coins—that can influence the outcome of the machine's digital prestidigitations.

Instead, to prevent their stock of coins from steadily draining away in repeated play, gamblers would have to count on machine failure or human error to destroy the illusion of randomness and remove some of the apparent unpredictability. An observant keno player once won $600,000 by noting that a particular casino computer that generated numbers for the game happened to restart its draw sequence whenever the power was turned off. He kept track of the numbers, identified the repeated pattern, and used it to score.

You can be sure, however, that slot-machine and casino operators watch their machines closely for any significant deviations from expected behavior, and anomalies come to light quickly. The odds in the games are rigged to favor the casinos, and casino operators vigilantly watch for anything that spoils the steady flow of coins into their coffers. Any windows of opportunity a gambler may see close very quickly when fortunes are at stake.

In situations in which we want an absolute assurance of unpredictability, as in an uncrackable cryptographic code (see chapter 6), a guarantee of pure randomness eludes us. Any suite of tests that we can apply—no matter how extensive—always leaves some degree of uncertainty. We can reduce the uncertainty by using multiple and varied tests, but a fragment of doubt always remains. Such is the slippery nature of randomness.

Scientifically, such niggling indeterminacy may not even matter much in the daily work of investigating the universe in which we live and in applying the knowledge we gain. "If the discipline you practice is sufficiently robust," the mathematician Mark Kac said in a 1983 essay on the meaning of randomness, "it contains enough checks and balances to keep you from committing errors that might come from the mistaken belief that you *really* know what 'random' is."

What we label as randomness represents, in some sense, a reflection of our ignorance and an acknowledgment of the limitations the world imposes on what we can know. At the same time, elusive randomness adds a necessary spice to our everyday lives, whether we call it luck, coincidence, or the wheel of fortune.

10

Lifetimes of Chance

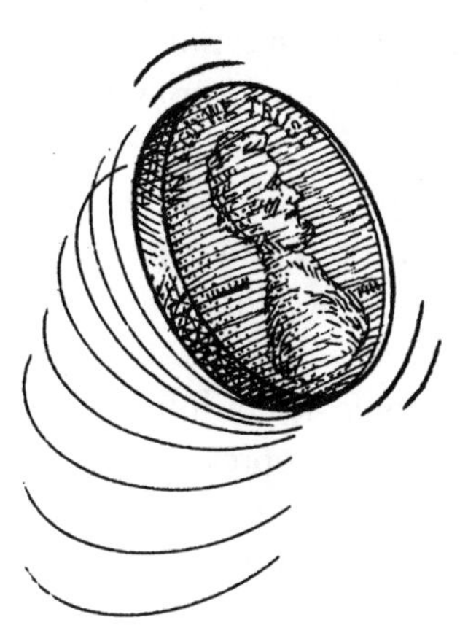

O! many a shaft at random sent
Finds mark the archer little meant!
And many a word, at random spoken,
May soothe or wound a heart that's broken!

—*Sir Walter Scott (1771–1832)*, "The Lord of the Isles"

The odds were fifty-five million to one, but the lines of people buying lottery tickets and dreaming of winning a $100 million jackpot seemed endless, snaking out of store doorways and down city blocks. For each dollar wagered, a bettor would simply pick five numbers from a pool of forty-five and an additional number from another group of forty-five.

Players had the choice of picking the numbers themselves or allowing the lottery computer to make a random selection for them. More often than not, they chose their own numbers, using various strategies to make a selection. Most people favored numbers based on the date of a special event—a birthday, a wedding anniversary, or the date of a divorce. Some looked to dreams, omens, and horoscopes for inspiration, and still others took their cues from the environment, noting the serial number on a dollar bill or the identifying number on a passing city bus. A few picked numbers by making star or triangle patterns on the lottery ticket.

In a study of lottery players, the sociologist Susan Losh of Florida State University found that some people even preferred to buy their own handheld calculatorlike random-number generators rather than rely on the lottery computer to make random selections for them. To Losh and others who have examined how people select lottery numbers, the strategies favored by the majority of players represent attempts to control chance and to impose some sort of structure on random events. The bettors exhibit a misplaced optimism that their efforts can somehow improve their chances of winning.

Actually, the most advantageous approach is to use a random-number generator—either the player's own or the lottery computer. That doesn't increase the chances of winning, but it does minimize the number of winning players who would have to split the pot. Because so many people use dates associated with commemorative events as the basis of their selections, certain sets of numbers tend be chosen more often than others. "Lucky" numbers, such as seven, also skew the distribution of selections, and when a winning combination happens to include a special date or number there are nearly always multiple winners.

Lotteries are not the only forum in which people try to structure a random process. In an effort to make life more predictable and more amenable to explanation, they invent theories and impose patterns on everyday events. Indeed, humans are predisposed to seeing order and meaning in a world that too often presents random choices and chaotic evidence.

Mathematics and statistics provide tools for sorting out which folk theories make sense and which represent nothing more than coincidence—the accidental concurrence of seemingly unrelated or random events. Similarly, the use of mathematics and statistics in science represents an attempt to establish cause-and-effect relationships, though rarely with absolute certainty, and to provide some form of rational explanation of the puzzles of our lives.

Amazing Coincidence

Nearly everyone has some sort of story to tell about a startling coincidence or the disproportionate effect of a seemingly inconsequential event, whether it is a missed airplane flight, a serendipitous encounter, or an odd mistake. Because of the role of chance in deciding whom we live with, work with, and marry, and in shaping many other facets of everyday life, our lives can appear to be driven by coincidence.

Nowadays newspapers, television broadcasts, Internet connections, and even our great mobility in driving around a neighborhood or hopping on a plane to travel across the country offer an extremely varied cross section of occurrences. Curious juxtapositions, accidental linkages, facts taken out of context, and seemingly miraculous events fuel the rationalizing coincidence engines of our minds. We are driven to invent story lines that link the disparate happenings of our lives.

▷ In the spring of 1996, the mathematics professor Celestino Mendez was discussing the statistical concept of expected value in his class at Metropolitan State College of Denver. He remarked that in a lottery the expected value—the probable gain or loss—increases as a jackpot gets progressively bigger. Mendez recommended that his students buy a ticket in the Colorado lottery, because the expected return at that time had risen to a gain of fourteen cents on each $1 ticket bought. (Normally, the expected return is nearly always negative because only about half the money collected is returned in the form of prizes, and players lose out in the long run.) On his way home that evening, the professor decided to take his own advice and bought ten tickets. He won, and shared the $15 million prize with one other player who had chosen the same numbers: 9, 12, 13, 14, 15, and 21.

- ▷ On June 29, 1990, the pitchers Dave Stewart of the Oakland Athletics and Fernando Valenzuela of the Los Angeles Dodgers hurled no-hitters. It was the first time in major-league-baseball history that a pitcher in each league on the same day had allowed no opposing players to hit the ball and get on base.
- ▷ In the issue that hit newsstands on April 7, 1912, the magazine *Popular Mechanics* carried a fictitious story about the maiden voyage of the largest ocean liner in the world. As the ship nears Newfoundland, an iceberg tears through the hull and she sinks. The story foreshadowed the ill-fated journey of the *Titanic*, which foundered on its first voyage after hitting an iceberg on the night of April 14, 1912.
- ▷ In the summer of 1979, a fifteen-year-old boy caught a ten-pound cod in a Norwegian fjord and presented the fish to his grandmother for preparation. When the grandmother opened up its stomach, she found inside a diamond ring—a family heirloom that she had lost while fishing in the fjord ten years earlier.
- ▷ The disks of the sun and the moon as seen from Earth are almost exactly the same size, giving us a particularly spectacular display of the sun's corona during a solar eclipse.

The mathematician Persi Diaconis and the statistician Frederick Mosteller of Harvard University have been pondering coincidences with the ultimate aim of developing a statistical theory to account for their occurrence. More than a decade ago, they began asking colleagues, friends, and friends of friends to send them examples of surprising coincidences. The collection has since ballooned into hundreds of file folders and dozens of notebooks filled with accounts.

Diaconis and Mosteller discovered that nearly all coincidences can be explained in terms of a few simple rules. "The world's activity and our labeling of events generates the coincidences," they contend. They found that they could classify their examples into several broad categories. Some coincidences have hidden causes, so they don't really count as coincidences. Others arise via psychological factors that make us unusually sensitive to certain chance events while ignoring others. So we think those events are unusual when in fact they are not, or we are inclined to perceive causal connections that are not really there.

Most coincidences, however, are simply chance events that turn out to be far more probable than many people imagine. "We are often surprised by things that turn out to be fairly likely occurrences," Diaconis and Mosteller note. One simple example of a coincidence that often surprises people involves birthdays. It's rather unlikely that you and I share the same birthday (month and date). The more people we pull into the group, however, the more likely it is that at least two people will have matching dates.

Ignoring the minor technicality of leap years, it's clear that in a group of 366 people, at least two must share a birthday. Yet it seems counterintuitive to many people that only twenty-three persons are needed in a group to have a fifty-fifty chance of at least one coincidental birthday. In other words, students in about half of all typically sized classrooms across the country are likely to find a shared birthday in the group.

To see why it takes only twenty-three people to reach even odds on sharing a birthday, we have to look at the probabilities. For one person, we can assume that all 365 days have an equal chance of being his or her birthday. So, the probability of one of those days being the birthday is 365/365. For a second person to have a birthday that doesn't match that of the first, he or she must be born on any one of the other 364 days of the year. One obtains the probability of no match between the birthdays of two people by multiplying $365/365 \times 364/365 = .9973$. Hence, the probability of a match is $1 - .9973$, or .0027, which is much less than 1 percent. For three people, the chance of having no pair of matching birthdays is $365/365 \times 364/365 \times 363/365 = .9918$.

As the number of people brought into the group increases, the chances of their being no match shrink. By the time the calculation reaches twenty-three people, the probability of no matching birthdays is .4927. Thus, the chances of at least one match within a group of twenty-three people is .5073, or slightly better than 50 percent. The reason the number of people is as low as twenty-three is that we aren't looking for a specific match. It doesn't have to be two particular people or a given date. Any match involving any date or any two people is enough to create a coincidence.

The same idea applies in many other situations. In 1986, for example, a New Jersey woman won a million-dollar lottery twice in four months. She was incredibly lucky, yet the odds of that happening to someone somewhere in the United States are only about one in thirty. Because so many people buy lottery tickets every day, someone among the millions of players is likely to win twice.

According to Diaconis, such outcomes illustrate the workings of what he calls the law of truly large numbers. "With a large enough sample, any outrageous thing is apt to happen," he says. He goes on to illustrate the point with the blade-of-grass paradox. Suppose someone stands somewhere in a field of grass and puts a finger on a single blade. The chances of choosing that particular blade may be millions to one, yet the probability of picking a blade of grass is 100 percent. So, if something happens to only one in a million people per day and the population of the United States is 250 million, "you expect 250 amazing coincidences every day," Diaconis says. Some fraction of these occurrences are bound to be reported somewhere in the press or remarked upon in the interminable discussions that take place on the Internet.

Another factor that sheds light on surprising coincidences is that many different chance events may qualify as noteworthy in a given situation. Two people meeting at a party might find that they share a birthday, come from the same town, have a friend in common, or have similar jobs. In truth, there are so many different things that two people could have in common that almost anything would count as a coincidence.

Another category of coincidences involves those requiring matches that are close but not exact. For example, it takes only fourteen people in a room to have even odds of finding two with birthdays that are identical or fall on consecutive days. Among seven people, there is about a 60 percent probability that two will have birthdays within a week of each other, and among four people the probability that two will have birthdays within thirty days of each other is about 70 percent. "Changing the conditions for a coincidence slightly can change the numbers a lot," Diaconis and Mosteller contend. To that end, "we find the utility of the birthday problem impressive as a tool for thinking about coincidences."

Such analyses are important not only for sorting out the meaning of coincidences in everyday life but also for revealing overlooked connections and hidden causes. When researchers find odd clusters of certain diseases, birth defects, or cancers in their data, statisticians have to decide whether these events represent the luck of the draw or may reflect an underlying cause. Similarly, in evaluating data from clinical trials of a new drug, they may have to determine whether a larger number of deaths among one particular subgroup of patients is a coincidence or means that the drug is dangerous to some people.

"Much of scientific discovery depends on finding the cause of a perplexing coincidence," Diaconis and Mosteller note in a paper on methods for studying coincidences. "At the same time, vast numbers of coincidences arise from hidden causes that are never discovered." Although coincidences truly reside in the mind of the observer, they drive much of our lives. "We are swimming in an ocean of coincidences," Diaconis and Mosteller conclude. "Our explanation is that *nature* and we ourselves are creating [them], sometimes causally, and also partly through perception and partly through objective accidental relationships. Often, of course, we cannot compute the probabilities, but when we can, such computations are informative."

Hot Hand

The star basketball player has the ball, but he has missed all five of his last shots. The television commentator wonders whether the player has gone cold and ought to be replaced. The player shoots and misses again. The coach signals from the sidelines.

Nearly everyone who plays or watches basketball has heard of a phenomenon known as a hot (or cold) hand. The term describes the supposed tendency of success (or failure) in basketball to be self-promoting or self-sustaining. On a good day, after making a couple of shots, players appear to relax, gain confidence, and slip into a groove, making success on subsequent shots more likely. "The belief in the hot hand . . . is really one version of a wider conviction that 'success breeds success' and 'failure breeds failure' in many walks of life," the psychologists Amos Tversky of Stanford University and Thomas Gilovich of Cornell University commented several years ago in a paper examining the statistics underlying the hot-hand phenomenon in basketball. "In certain domains, it surely does—particularly those in which a person's reputation can play a decisive role," they continued. "However, there are other areas, such as most gambling games, in which the belief can be just as strongly held, but where the phenomenon clearly does not exist."

Where does basketball's hot hand fit in? Is success in basketball self-promoting, so that players shoot in streaks, or are the chances of hitting a shot as good after a miss as after a hit? Consider the following question: What are the chances that a basketball player who averages 47 percent success in shooting will miss eight of the next ten shots? One could try to answer the question by looking up a formula in a reference book and plugging in numbers to calculate a probability. Alternatively, one can perform a simple numerical experiment to come up with a useful result.

The computer generates ten random numbers between 1 and 100. It counts all the numbers that fall between 0 and 53 as missed shots. It then spews out a second sequence of ten random numbers, and so on, each time counting up the number of misses. After repeating the process, say, a hundred times, the computer determines the number of trials in which the player misses eight or more shots. This figure establishes the likelihood of the player's suffering a streak of misses, showing in this case that the player will miss eight or more shots out of ten tries about 6.5 percent of the time.

Because of random variation, such streaks of misses occur naturally, even when a player is not in a slump. Hence, when a good player misses eight out of ten shots, it doesn't necessarily mean that he or she must be replaced. The streak could quite easily occur by chance, just as a string of eight heads in a row has a reasonable probability of showing up in 125 tosses of a coin (see chapter 1). People tend to misjudge how often such streaks can happen by chance.

In their study of basketball's hot hand, Tversky and Gilovich found no evidence of the hot-hand phenomenon in any of their data sets, which included shots taken from the field and pairs of free throws during National Basketball Association games, as well as a controlled experiment involving college men and women basketball players. "Our research does not indicate that basketball shooting is a purely chance process, like coin tossing. Obviously, it requires a great deal of talent and skill," the two psychologists concluded. "What we have found is that, contrary to common belief, a player's chances of hitting are largely independent of the outcome of his or her previous shots." Naturally, a player may now and then make, say, nine of ten shots. An observer may then claim, after the fact, that the player was hot. "Such use, however, is misleading if the length and frequency of such streaks do not exceed chance expectation," Tversky and Gilovich argued.

Of course, such factors as the willingness of a player to shoot may be affected by his perception of success, and, in the long run, a player's performance may improve or worsen overall. Short-run analyses of day-to-day and week-to-week individual and team performances in most sports, however, show results similar to the outcomes that a good random-number generator would produce.

Tversky, who died in 1996, was fond of describing his work as "debugging human intuition." With his colleague Daniel Kahnemann, he conducted a series of studies of the way people judge probabilities and estimate the likely outcome of events. As in the hot-hand example, Tversky could establish again and again the existence of mismatches between intuition and probability—between cognitive illusion and reality.

The human tendency to look for order and to spot patterns is enormously helpful in everyday life, especially when such hunches are subject to further tests. Trouble brews when we lower our standards of evidence and treat these hypotheses as established facts.

Pick a Sample

Belief in nonexistent patterns is costly, says the economist Julian L. Simon of the University of Maryland. Scores of investors pay a high price for information that purports to predict stock market performance based on finding patterns in the rise and fall of stock prices. The results are often no better than those achieved by throwing darts at a stock listing.

Several decades ago, Simon conducted an experiment in which he cooked up a set of random numbers that he described as the weekly prices of publicly traded stocks. Participants were told to buy and sell stocks according to the quoted prices. Each week for ten weeks, he gave them a new set of numbers. "The players did all sorts of fancy calculating, using a wild variety of assumptions—although there was no possible way that the figuring could help them," Simon says. The game was so compelling, however, that players wanted to continue, even after they were told partway through that because the prices were randomly generated numbers, winning or losing didn't depend on their skills.

Such experiences also made Simon aware of the trouble that many people have understanding and using statistics. The issue first struck home in 1967 when he realized that all four of the graduate students who were taking his course on research methods in business had used "wildly wrong" statistical tests to analyze data for their class projects. "The students had swallowed but not digested a bundle of statistical ideas—which now misled them—taught by professors who valued fancy mathematics even if useless or wrong," Simon says.

The experience led Simon to develop an alternative approach to teaching statistics. His method emphasizes intuitive reasoning and involves using a computer to perform experiments on available data—a process known as resampling—to get meaningful answers to statistical problems. In his scheme, students avoid having to tangle with mysterious formulas, cryptic tables, and other forms of mathematical magic to get their results.

In a sense, the resampling approach brings statistics back to its roots. More than three hundred years ago, before the invention of probability theory, gamblers had to rely on experimentation—by dealing hands or throwing dice—to estimate odds. The computer has allowed the return of an empirical approach to the center of statistics, shifting emphasis away from the formula-based statistical methods developed between 1800 and 1930, when computation was slow and expensive and had to be avoided as much as possible.

A resampling experiment uses a mechanism such as a coin, a die, a deck of cards, or a computer's stock of random numbers to generate sets of data that represent samples taken from some population. The availability of such "artificial" samples permits students—as well as researchers—to experiment systematically with data in order to estimate probabilities and make statistical inferences.

The resampling approach addresses a key problem in statistics: how to infer the "truth" from a sample of data that may be incomplete or drawn from an ill-defined population. For example, suppose ten measurements of some quantity yield ten slightly different values. What is the best estimate of the true value?

Such problems can be solved not only by the use of randomly generated numbers but also by the repeated use of a given set of data. Consider, for instance, the dilemma faced by a company investigating the effectiveness of an additive that seems to increase the lifetime of a battery. One set of experiments shows that ten batteries with the additive lasted an average of 29.5 weeks and that ten batteries without the additive lasted an average of 28.2 weeks. Should the company take the difference between the averages seriously and invest substantial funds in the development of additive-laced batteries?

To test the hypothesis that the observed difference may be simply a matter of chance and that the additive really makes no difference, the first step involves combining the two samples, each of which contains ten results, into one group. Next, the combined data are thoroughly mixed, and two new samples of ten each are drawn from the shuffled data. The computer calculates the new averages and determines their difference, then repeats the process a hundred or more times.

The company's researchers can examine the results of this computer simulation to get an idea of how often a difference of 1.3 weeks comes up when variations are due entirely to chance. By observing the possible variation from sample to sample, they get a sense of how variable the data can be. That information helps them evaluate the additive's true efficacy.

In 1977, Bradley Efron, a statistician at Stanford University, independently developed a resampling technique for handling a variety of complex statistical problems. The development and application of that technique, called the "bootstrap," has given Simon's approach a strong theoretical underpinning.

In essence, the bootstrap method substitutes a huge number of simple calculations, performed by a computer, for the complicated mathematical formulas that play such a key role in conventional statistical theory. It provides a way of using the data contained in a single sample to estimate the accuracy and variability of the sample average or some other statistical measure. The idea is to extract as much information as possible from the data on hand. Suppose a researcher has only a single data set, consisting of fifteen numbers. In the bootstrap procedure the computer copies each of these values, say, a billion times, mixes them thoroughly, then selects samples consisting of fifteen values each. The samples can then be used to estimate, for example, the true average and to determine the degree of variability in the data.

The term bootstrap—derived from the old saying about pulling yourself up by your own bootstraps—reflects the fact that the one available sample gives rise to all the others. Such a computationally intensive approach provides freedom from two limiting factors that dominate traditional statistical theory: the assumption that the data conform to a bell-shaped curve, called the normal or Gaussian distribution; and the need to focus on statistical measures whose theoretical properties can be analyzed mathematically. "Part of my work is just trying to break down the prejudice that the classical way is the only way to go," Efron says. "Very simple theoretical ideas can get you a long way. You can figure out a pretty good answer without going into the depths of [classical] theory."

The bootstrap approach is one of a number of mathematical and statistical techniques that help shed light on the beliefs that we hold and point to the need to look critically at the data we have in hand. The techniques serve as guides for determining what is worth believing and what should be regarded skeptically. Instead of absolute certainty, however, they offer only glimpses of probable truths.

Random Reality

In contemplating the toss of a coin, the output of a computer's random-number generator, the wild movements of an amusement park ride, the decay of a radioactive atom, the erratic beating of a failing heart, the path of a perfume molecule drifting in the air, and the varied byways of coincidence, one might conclude that there are several different kinds of randomness. In many cases, the system is so complicated that no one can untangle the precise causes of its behavior. In others, the underlying mechanism is simple and deterministic, but the results look highly unpredictable. In some situations, quantum uncertainty at the microscopic level yields an intrinsic randomness that cannot be avoided.

We use mathematics and statistics to describe the diverse realms of randomness. From these descriptions, we attempt to glean insights into the workings of chance and to search for hidden causes. With such tools in hand, we seek patterns and relationships and propose predictions that help us make sense of the world.

When we turn to mathematics for answers, however, we also confront the limitations of mathematics itself. Suppose you happen upon the number 1.6180339887. It looks vaguely familiar, but you can't quite place it. You would like to find out whether this particular sequence of digits is special in some way, perhaps as the output of a specific formula or the value of a familiar mathematical expression. In other words, is there a hidden pattern?

It turns out that this number is the value, rounded off, of the expression $(1 + \sqrt{5})/2$, which represents the golden ratio of the sides of a rectangle, often found in Greek architecture and design. Given the formula, it's a straightforward task to compute the value of this ratio to any number of decimal places. Going in the opposite direction from the number to the formula, however, is much more difficult and problematic. For example, it's possible that if the mystery number were available to an extra decimal place, the final digit would no longer match the output of the formula because the formula really doesn't apply. Moreover, there may be several formulas that give the same sequence of digits. Or perhaps there is no formula at all, and the digits merely represent a random sequence.

Like the digits of the golden ratio, the digits of π (pi) also continue in an endless, never-repeating chain of numbers. One intriguing issue is whether there is any pattern among the digits of π or whether the digits could pass for a sequence of random numbers. Of course, the crucial difference between the digits of π and a sequence of random numbers generated by tossing a perfectly balanced coin is that π can be produced by a simple mathematical operation: dividing a circle's circumference by its diameter.

In other words, the infinite digits of π can be compressed into a straightforward formula. Similarly, the sequence of numbers 1, 2, 4, 8, 16, 32, 64, 128, . . . and the sequence of digits 1001001001001001 . . . can both be compressed into simple rules or short computer programs for generating the string of numbers.

In contrast, suppose we are given the following string of digits derived from tossing a coin (1 for heads, 0 for tails): 11100101001110. Unless we accidentally hit upon the digits of some well-known expression (and the chances of that happening are infinitesimally small), we have no shorthand way of writing out the sequence other than by listing all the digits. One could say that a random string of numbers is its own shortest description. In a sense, it takes vastly more information to express a random sequence than to express the digits of π.

In order to sort through the relationship between random sequences and the types of numbers that mathematicians and scientists use in their work, Gregory J. Chaitin, a mathematician at IBM Research, has defined the "complexity" of a number as the length of the shortest computer program (or set of instructions) that will spew out the number. Chaitin's work on what he calls algorithmic information theory extends and refines ideas independently introduced in the 1960s by the Russian scientist Andrey N. Kolmogorov (1903–1987) and others.

Suppose the given number consists of one hundred 1s. The instruction to the computer would simply be to "print 1, 100 times." Because the program is substantially shorter than the sequence of a hundred 1s that it generates, the sequence is not considered random. Putting in instructions for printing out a sequence of alternating 1s and 0s lengthens the program only slightly, so that sequence also isn't random.

If the sequence is disorderly enough that any program for printing it out cannot be much shorter than the sequence itself, then the sequence counts as random. Hence, according to Chaitin and Kolmogorov's definition, a random sequence is one for which there is no compact description.

Chaitin proved that no program can generate a number more complex than itself. In other words, "a one-pound theory can no more produce a ten-pound theorem than a one-hundred-pound pregnant woman can birth a two-hundred-pound child," Chaitin likes to say. Conversely, he also showed that it is impossible for a program to prove that a number more complex than the program is random. Hence, to the extent that the human mind is a kind of computer, there may be a type of complexity so deep and subtle that our intellect could never grasp it. Whatever order may lie in the depths would be inaccessible, and it would always appear to us as randomness.

At the same time, proving that a sequence is random also presents insurmountable difficulties. There's no way to be sure that we haven't overlooked a hint of order that would allow even a small compression in the computer program that reproduces the sequence. Whether the numbers come from the toss of a coin or from static picked up by a radio telescope, we can never know whether we have extracted the right message from the data, assuming there is a message in the first place.

From a mathematical point of view, Chaitin's result also suggests that we are far more likely to find randomness than order within certain domains of mathematics itself. For example, one can imagine labeling the points that make up a line segment with numbers. Suppose the two ends of the line are marked 0 and 1. The interval between 0 and 1 contains an infinite number of points, each one specified by a number drawn from an infinite collection of decimals.

Some of the numbers are fractions, starting with ½ or 0.5 in the middle, then ¼ (0.25) and ¾ (0.75) dividing the two halves of the line, and so on. The line, in fact, encompasses all possible fractions, which can be expressed as decimals of a finite length or as infinitely long strings in which a group of digits repeats periodically. That still leaves a host of numbers, such as the decimal digits of π, which we can express, but not as fractions. Beyond that are infinitely long numbers for which we can find no expression. Such irrational numbers vastly outnumber the fractions, which means that all but a tiny, insignificant proportion of the numbers labeling the points of a line segment are random strings. Hence, nearly every number is random, yet we can't count on proving that any individual number is truly random.

The implications of a direct connection between determinism, logic, and randomness are startling. In 1931, the mathematician Kurt Gödel (1906–1978) proved that any mathematical system of sufficient power, such as the whole numbers and the ordinary arithmetic that we use in everyday life, contains statements that cannot be proved true or false within the given system of axioms and rules used for deducing results about those statements to generate theorems. That discovery came as an astonishing, unexpected blow to mathematicians schooled in a tradition that went back two thousand years to the Greek geometer Euclid.

In Euclid's world, one could start with a short list of simple axioms and certain logical rules and then prove other properties of triangles, numbers, or whatever geometric objects were of interest. What could be more certain than the fact that $2 + 2 = 4$, or that any angle inscribed in a semicircle is a right angle?

The general belief was that one could derive all of mathematics from an appropriate set of axioms. Indeed, at the beginning of the twentieth century the German mathematician David Hilbert declared that all mathematical questions are in principle decidable—that is, they can be shown to be either true or false. He confidently set out to codify methods of mathematical reasoning to demonstrate that any precisely stated mathematical question would have a clear resolution via a mathematical proof.

In essence, given a well-defined set of axioms and appropriate rules of inference, his idea was to run through all possible proofs, starting with the shortest, and check which ones are correct. It's like having millions of monkeys typing randomly and generating the world's literature, except that here, they would be writing mathematical expressions. In principle, Hilbert's procedure would generate all possible theorems.

In Gödel's world, no matter what the system of axioms or rules is, there will always be some property that can be neither proved nor invalidated within the system. For example, there could be mathematical propositions about integers that are true but can only be noted and never proved. Indeed, mathematics is full of conjectures—questions waiting for answers—with no assurance that the answers even exist.

Mathematicians have sometimes assumed the truth of unproved but useful conjectures in their work, occasionally relying on evidence from extensive computer calculations to bolster their confidence that the assertion is likely to be true. In some situations, such testing rather than a rigorous mathematical proof may be all that anyone can count on. Number theory becomes an experimental science instead of an exercise in logic.

Chaitin's work goes further in suggesting that there is an infinite number of mathematical statements that one can make about arithmetic that can't be reduced to the axioms of arithmetic, so there's no way to prove whether the statements are true or false by using arithmetic. In Chaitin's view, that's practically the same as saying that the structure of arithmetic is random. "What I've constructed and exhibited are mathematical facts that are true . . . by accident," he says. "They're mathematical facts which are analogous to the outcome of a coin toss. . . . You can never actually prove logically whether they're true."

This doesn't mean that anarchy reigns in mathematics, only that mathematical laws of a different kind might apply in certain situations. In such cases, statistical laws hold sway and probabilities describe the answers that come out of the equations. Such problems arise when one asks whether an equation involving only whole numbers has an infinite number of whole-number solutions, a finite number, or none at all. "In the same way that it is impossible to predict the exact moment at which an individual atom undergoes radioactive decay, mathematics is sometimes powerless to answer particular questions," Chaitin states. "Nevertheless, physicists can still make reliable predictions about averages over large ensembles of atoms. Mathematicians may in some cases be limited to a similar approach."

The late physicist Joseph Ford spent a great deal of time and effort probing the implications of Chaitin's discovery for science and for what we can learn about the world in which we live. Scientists typically assume that if they pose their questions appropriately, nature provides only one, unambiguous answer to each query and the result is part of some complete, overarching theory, which it is their duty to unearth.

What's the connection between Chaitin's definition of randomness and the physical world? Imagine reducing the description of any physical system to a sequence of digits. For example, you could express the positions and velocities of all the gas molecules bouncing around inside a balloon as strings of 1s and 0s. The idea is to regard such a physical system's initial configuration as input information for a computer simulation of its future behavior.

In ordinary systems, a small amount of information is converted into a large quantity of output information in the form of reasonably accurate predictions. For random systems, such a simulation is pointless. The computer is reduced to being merely a copying machine. At best, the system generates as much information as the amount put in. The computer can describe the system in increasingly greater detail, but it can't make any worthwhile predictions.

As Gödel showed, the mathematical logic that serves as the language and foundation of science is incomplete. And no one has yet proved that the axioms we work with every day in mathematics are consistent. Thus, there's no guarantee that a given assertion can be proved true or false, and no guarantee that a result and its opposite can't both be proved correct. So, it isn't known with certainty that 2 + 2 doesn't give answers other than 4. Conceivably, if the axioms of mathematics turn out to be inconsistent, you could prove that 2 + 2 = 4 and that 2 + 2 = 3. Once you do that, the system falls apart, and you can prove anything.

There's a cartoon that shows a mathematician staring quizzically at a simple sum scrawled on a blackboard. The caption reads, "The arithmetic seems correct, yet I find myself haunted by the idea that the basic axioms on which arithmetic is based might give rise to contradictions that would then invalidate these computations."

In mathematics, there will always be questions that can be asked but not answered in any particular system of logic. Ironically, such hard-to-answer questions are often easy to ask, and there's really no way to tell in advance, either in mathematics or science, whether the effort to search for a proof or an explanation will be rewarded with an answer. Sometimes, though, the effort is its own reward.

"The randomness which lies at the very foundations of pure mathematics of necessity permeates every human description of nature," Ford contends. In his view, the randomness of coin tosses, chaotic weather systems, and quantum processes has a single face that reflects the limitations of the mathematics that we use in attempting to predict the universe.

Blind Chance

In the early nineteenth century, the German philosopher Georg Hegel (1770–1831) offered the image of an enveloping absolute spirit at the center of the universe that guides all reality. "The aim of science is to reduce the scope of chance," he insisted. His vision was one of a world functioning according to its own rules, which we, at least in principle, could discover and make use of.

Those patterns, rules, and laws that we have teased out of the jungles of randomness have provided the means by which we have survived and prospered as a species. In our theories, we have constructed oases of regularity and told stories of how we came to be. The universe, however, is not an open book, with the beginning, middle, and end laid out before us. Our efforts to understand nature invariably fall short, and we are left with fragments of information and only a partial understanding. Even our mathematical tools have limitations.

The statistician David Blackwell of the University of California at Berkeley once suggested that most of mathematics may be chaotic, and that it is only the small part where we recognize patterns that we actually call mathematics. The mathematical study of combinatorial games tends to foster such a view, says the mathematician Richard K. Guy of the University of Calgary. "Are games with complete information and no chance moves of little interest because there are pure winning strategies?" he asks. "On the contrary, when we come to analyze even the simplest of such games, we often run into regions of complete chaos."

Randomness, chaos, uncertainty, and chance are all a part of our lives. They reside at the ill-defined boundaries between what we know, what we can know, and what is beyond our knowing. They make life interesting. We are fortunate to live on a planet in a solar system in a galaxy in a universe in which the balance between stasis and complete and utter disorder favors a restless mind—one that can take advantage of pockets of order and pattern to ensure survival yet also learn from the surprises that chance offers. We can theorize ad infinitum. We can make choices. We can build a machine as complex as a computer, which behaves so predictably over such long periods that we can rely upon its calculations most of the time. We can use probabilities, statistical methods, and even randomness itself to shed light on issues that perplex us.

The results, however, are always conditional, and absolute certainty eludes us. There are fundamental limits to what we can know, what we can foretell, and how much we can understand of both the physical universe and the universe of mathematical experience. What we cannot attain we label as random, or unknowable. But perhaps randomness itself is a crucial unifying principle. We see it in quantum mechanics and in the behavior of atoms, in the classical physics of billiard balls, vibrating drums, tossed coins, and weather systems, and even among the whole numbers of pure mathematics and elementary number theory.

Just as one can't predict the outcome of the next toss in a series of coin flips, one can't necessarily answer a physical or a mathematical question in a particular case. However, just as one can determine that in the long run heads come up about half the time, one can also obtain statistical results that tell us something about the physical or mathematical behavior in question. The mathematics we use in the gambling casino also applies to much of the rest of the universe!

We are surrounded by jungles of randomness. With our mathematical and statistical machetes, we can hack out extensive networks of trails and clearings that provide for most of our day-to-day needs and make sense of some fraction of human experience. The vast jungle, however, remains close at hand, never to be taken for granted, never to reveal all its secrets—and always teasing the inquiring mind.

Appendix

Secret Messages

To generate a cryptogram, the sender first rewrites the message as a string of binary numbers according to a conversion table (A = 01000001, B = 01000010, and so on). This string is then added to an equally long sequence of random bits (the secret key) using the basic rules of binary modular arithmetic:

$$0 + 0 = 0,$$

$$0 + 1 = 1 + 0 = 1,$$

$$1 + 1 = 0.$$

The sender transmits the resulting sequence, which is the encrypted message.

Message	*H*	*E*	*L*	*P*
Message (in binary)	01001000	01000101	01001100	01010000
Secret key	10100010	10101010	10100101	01110101
Encrypted message	11101010	11101111	11101001	00100101
Decrypted message	01001000	01000101	01001100	01010000

Using the same secret key, the recipient recovers the original digital text by adding the digits of the key (following the same rules as above) to the transmitted sequence.

Riddled Basins

Modern computer graphics techniques reveals an astonishingly rich landscape of filigreed features lurking within modest mathematical expressions.

Consider points defined by two coordinates, x and y. For a given starting point (x, y), one can compute a new value of x by using the formula $x^2 - y^2 - x - ay$, and a new value of y by using $2xy - ax + y$. The parameter a can take on various values, but Alexander and his colleagues used a special value, $a = 1.0287137\ldots$, which makes the resulting orbits somewhat easier to characterize.

The next step is to compute systematically the trajectories of a bunch of starting points, for which x and y range between 2 and -2. One can then identify on which one of three attractors a point eventually lands and color-code each starting point accordingly. Different shades of a color correspond to how rapidly a given starting point converges on a certain attractor.

The procedure requires a little bit of bookkeeping, but in principle the calculations and colorings are extremely easy to do. The resulting image, with three different colors used to represent starting points that end up on each of the three attractors, clearly shows the intricacy of the meshed basins belonging to the attractors. Voilà! A foamy portrait of intimately intermingled basins.

Random Numbers

In a typical linear congruential generator, a starting number of n digits is multiplied by some constant r. Another constant c is added to the result, and the answer is then divided by a number m. The remainder after division constitutes the required random number. It also becomes the seed for the next round of steps. With a proper choice of multiplier and divisor, such a generator produces a sequence of numbers that are difficult to distinguish from random numbers generated purely by chance.

Researchers have focused a great deal of attention on selecting the best possible values for r, c, and m. When m is a power of 2, for example, a suitable sequence can be generated if, among other things, c is odd, and $r - 8$ is a multiple of 8. A simple example of a five-bit number sequence using these conditions is to set $m = 2^5$ (or 32), $r = 21$, and $c = 1$. Starting with the seed number 13, these specifications yield the sequence 13, 18, 27, 24, 25, 14, 7, 20, 5, 10. . . . Written in binary form, the sequence consists of the terms 01101, 10010, 11011, 11000, 11001, 01110, 00111, 10100, 00101, 01010. . . .

Such a sequence looks quite random. Notice, however, that the numbers alternate between odd and even, so in the binary representation, the final bit alternates between 1 and 0. Moreover, the second bit from the right in successive numbers follows the repeated pattern

0110. The third bit from the right has a pattern that repeats after eight digits, and so on.

Nonetheless, one can diminish the influence of such patterns by going to much larger numbers and different values of the constants. One widely used variant requires the values $r = 16{,}807$, $c = 0$, and $m = 2^{31} - 1$.

Fibonacci Randomness

At the heart of the new Marsaglia-Zaman method of generating random numbers is the so-called Fibonacci sequence. The classic example of a Fibonacci sequence consists of the numbers 1, 1, 2, 3, 5, 8, 13, 21, 34, 55, 89. . . . Except for the first 1, each number in the sequence equals the sum of the previous two numbers. For example, 89 = 34 + 55. The next number of the sequence would be 55 + 89, or 144.

Taking only the last digit of each number in the sequence yields the numbers 1, 1, 2, 3, 5, 8, 3, 1, 4, 5, 9, 4. . . . However, at the sixty-first term, the sequence of final digits starts to repeat itself.

To extend such a generator's range of usable numbers, Marsaglia and Zaman made a simple but crucial modification. They introduced a "carry bit," which may have an initial value of either 0 or 1. The resulting "add-with-carry" generator then has some of the features associated with "long addition," an operation familiar to those who learned arithmetic in the days before calculators.

In the Marsaglia-Zaman procedure, suppose the seed value is 1 and the initial carry bit is 0. Each new number in the sequence equals the sum of the previous two numbers plus the carry bit. If the result is greater than 10, the carry bit used in computing the next number in the sequence becomes 1. The result is the following Fibonacci-like sequence: 1, 1, 2, 3, 5, 8, 13, 22, 36, 58, 94, 153. . . . A sequence of pseudorandom numbers can then be derived from the final digits of these numbers: 1, 1, 2, 3, 5, 8, 3, 2, 6, 8, 4, 3. . . . Now, the sequence goes on for 108 digits before it repeats.

By selecting appropriate starting values and slightly altering the arithmetic, computer programmers can create a whole class of add-with-carry and related generators that have periods considerably longer than those of the conventional generators now in use.

Bibliography

General

Devlin, Keith. 1994. *Mathematics: The Science of Patterns*. New York: Scientific American Library.

Ekeland, Ivar. 1993. *The Broken Dice and Other Mathematical Tales of Chance*. Chicago: University of Chicago Press.

Peterson, Ivars. 1988. *The Mathematical Tourist: Snapshots of Modern Mathematics*. New York: W. H. Freeman.

Stewart, Ian. 1996. *From Here to Infinity: A Guide to Today's Mathematics*. Oxford, England: Oxford University Press.

Preface: Infinite Possibility

Carroll, Lewis. 1990. *More Annotated Alice: Alice's Adventures in Wonderland and Through the Looking-Glass* (notes by Martin Gardner). New York: Random House.

Huff, Darrell, and Irving Geis. 1959. *How to Take a Chance*. New York: W. W. Norton.

Kac, Mark. 1983. What is random? *American Scientist* (July/August): 405–406.

Maloney, Russell. 1940. Inflexible logic. In *The World of Mathematics*. Vol. 4, James R. Newman, ed. New York: Simon & Schuster.

Monod, Jacques. 1972. *Chance and Necessity: An Essay on the Natural Philosophy of Modern Biology*. New York: Random House.

1. The Die Is Cast

Albers, Donald J., Gerald L. Alexanderson, and Constance Reid, eds. 1990. *More Mathematical People: Contemporary Conversations*. San Diego, Calif.: Academic Press.

Beasley, John D. 1989. *The Mathematics of Games*. Oxford, England: Oxford University Press.

Broline, Duane M. 1979. Renumbering of the faces of dice. *Mathematics Magazine* 52 (November): 312–315.

Daston, Lorraine. 1988. *Classical Probability in the Enlightenment.* Princeton, N.J.: Princeton University Press.

Diaconis, Persi, and Joseph B. Keller. 1989. Fair dice. *American Mathematical Monthly* (April): 337–339.

Dostoyevsky, Fyodor. 1996. *The Gambler.* New York: Dover.

Gallian, Joseph. 1995. Weird dice. *Math Horizons* (February): 30–31.

Gallian, Joseph A., and David J. Rusin. 1979. Cyclotomic polynomials and nonstandard dice. *Discrete Mathematics* 27: 245–259.

Gardner, Martin. 1989. *Mathematical Magic Show.* Washington, D.C.: Mathematical Association of America.

——. 1978. Mathematical games: On checker jumping, the amazon game, weird dice, card tricks and other playful pastimes. *Scientific American* 238: 19–32.

Gigerenzer, Gerd, et al. 1989. *The Empire of Chance: How Probability Changed Science and Everyday Life.* Cambridge, England: Cambridge University Press.

Grunfeld, Frederic V., ed. 1975. *Games of the World.* New York: Holt, Rinehart & Winston.

Hacking, Ian. 1975. *The Emergence of Probability: A Philosophical Study of Early Ideas about Probability, Induction and Statistical Inference.* Cambridge, England: Cambridge University Press.

Huff, Darrell, and Irving Geis. 1959. *How to Take a Chance.* New York: W. W. Norton.

Orkin, Mike. 1991. *Can You Win? The Real Odds for Casino Gambling, Sports Betting, and Lotteries.* New York: W. H. Freeman.

Mohr, Merilyn Simonds. 1993. *The Games Treasury.* Shelburne, Vt.: Chapters Publishing.

Packel, Edward. 1981. *The Mathematics of Games and Gambling.* Washington, D.C.: Mathematical Association of America.

Scarne, John. 1986. *Scarne's New Complete Guide to Gambling.* New York: Simon & Schuster.

Schilling, Mark F. 1994. Long run predictions. *Math Horizons* (spring): 10–12.

Stewart, Ian. 1996. Monopoly revisited. *Scientific American* 274 (October): 116–119.

——. 1996. How fair is Monopoly? *Scientific American* 274 (April): 104–105.

Vulovic, Vladimir Z., and Robert E. Prange. 1986. Randomness of a true coin toss. *Physical Review* A 33 (January): 576–582.

Weaver, Warren. 1963. *Lady Luck: The Theory of Probability.* Garden City, N.Y.: Doubleday.

Wykes, Alan. 1964. *The Complete Illustrated Guide to Gambling.* Garden City, N.Y.: Doubleday.

2. Sea of Life

Albertson, Michael O. 1994. People who know people. *Mathematics Magazine* 67 (October): 278–281.

Cipra, Barry. 1995. A visit to Asymptopia yields insight into set structures. *Science* 267 (17 February): 964–965.

Gardner, Martin. 1989. *Penrose Tiles to Trapdoor Ciphers.* New York: W. H. Freeman.

Graham, R. L., and J. Nesetril. 1996. *The Mathematics of Paul Erdos I.* New York: Springer-Verlag.

Graham, Ronald L., Bruce L. Rothschild, and Joel H. Spencer. 1990. *Ramsey Theory,* 2nd ed. New York: Wiley.

Graham, Ronald L., and Joel H. Spencer. 1990. Ramsey theory. *Scientific American* (July): 112–117.

Hoffman, Paul. 1987. The man who loves only numbers. *Atlantic Monthly* 260 (November): 60–74.

Kolata, Gina. 1984. Order out of chaos in computers. *Science* 223 (2 March): 917–919.

Odda, Tom (Ronald L. Graham). 1979. On properties of a well-known graph, or what is your Ramsey number? *Annals of the New York Academy of Sciences* 328 (20 June): 166–172.

Peterson, Ivars. 1993. Party numbers. *Science News* 144 (17 July): 46–47.

Russell, Bertrand. 1996. *The Principles of Mathematics.* New York: W. W. Norton.

Tierney, John. 1984. Paul Erdos is in town. His brain is open. *Science* 84 (October): 40–47.

3. Shell Game

Berger, Bonnie, et al. 1994. Local rule-based theory of virus shell assembly. *Proceedings of the National Academy of Sciences USA* 91 (August): 7732–7736.

Dutton, Gail. 1996. CAT antiviral agents block the formation of virus shell. *Genetic Engineering News* (15 November): 1.

Fuller, R. Buckminster. 1979. *Synergetics 2: Further Explorations in the Geometry of Thinking.* New York: Macmillan.

——. 1975. *Synergetics: Explorations in the Geometry of Thinking.* New York: Macmillan.

Kardar, Mehran. 1996. Which came first: Protein sequence or structure? *Science* 273 (2 August): 610.

Lander, Eric S., and Michael S. Waterman, eds. 1995. *Calculating the Secrets of Life.* Washington, D.C.: National Academy Press.

Levine, Arnold J. 1992. *Viruses.* New York: Scientific American Library.

Li, Hao, et al. 1996. Emergence of preferred structures in a simple model of protein folding. *Science* 273 (2 August): 666–669.

Nesse, Randolph M., and George C. Williams. 1994. *Why We Get Sick: The New Science of Darwinian Medicine.* New York: Times Books.

Peterson, Ivars. 1995. Viral shell game. *Science News* 147 (25 March): 186–187.

Prevelige, Peter E., Jr., Dennis Thomas, and Jonathan King. 1993. Nucleation and growth phases in the polymerization of coat and scaffolding subunits into icosahedral procapsid shells. *Journal of Biophysics* 64 (March): 824–835.

Radetsky, Peter. 1991. *The Invisible Invaders: Viruses and the Scientists Who Pursue Them.* New York: Little, Brown & Company.

Richards, Frederic M. 1991. The protein folding problem. *Scientific American* (January): 54–63.

Stewart, Ian. 1989. *Game, Set, and Math: Enigmas and Conundrums.* Oxford, England: Basil Blackwell.

Wolynes, Peter G. 1996. Symmetry and the energy landscapes of biomolecules. *Proceedings of the National Academy of Sciences USA* 93 (December): 14249–14255.

4. Call of the Firefly

Buck, John. 1988. Synchronous rhythmic flashing of fireflies II. *Quarterly Review of Biology* 63: 265–289.

——. 1938. Synchronous rhythmic flashing of fireflies. *Quarterly Review of Biology* 13: 301–314.

Buck, John, and Elisabeth Buck. 1976. Synchronous fireflies. *Scientific American* (May): 74–85.

Cohen, Avis H., et al. 1992. Modelling of intersegmental coordination in the lamprey central pattern generator for locomotion. *Trends in Neurosciences* 15 (11): 434–438.

Collins, Jim, and Ian Stewart. 1994. The mathematical springs in insect steps. *New Scientist* (8 October): 36–40.

Collins, J. J., and I. N. Stewart. 1993. Coupled nonlinear oscillators and the symmetries of animal gaits. *Journal of Nonlinear Science* 3 (July): 349–392.

Ermentrout, G. B., and N. Kopell. 1994. Inhibition-produced patterning in chains of coupled nonlinear oscillators. *SIAM Journal on Applied Mathematics* 54 (April): 478–507.

———. 1990. Phase transitions and other phenomena in chains of coupled oscillators. *SIAM Journal on Applied Mathematics* 50 (August): 1014–1052.

———. 1990. Oscillator death in systems of coupled neural oscillators. *SIAM Journal on Applied Mathematics* 50 (February): 125–146.

Garver, Wayne, and Frank Moss. 1993. Electronic fireflies. *Scientific American* (December): 128–130.

Grillner, Sten. 1996. Neural networks for vertebrate locomotion. *Scientific American* (January): 64–69.

Hanson, Frank E. 1982. Pacemaker control of rhythmic flashing of fireflies. *Cellular Pacemakers* 2: 81–100.

Hanson, Frank E., et al. 1971. Synchrony and flash entrainment in a New Guinea firefly. *Science* 174 (8 October): 161–164.

Jackson, Allyn. 1991. Lamprey lingo. *Notices of the American Mathematical Society* 38 (December): 1236–1239.

Koch, Christof. 1997. Computation and the single neuron. *Nature* 385 (16 January): 207–210.

Kopell, N., G. B. Ermentrout, and T. L. Williams. 1991. On chains of oscillators forced at one end. *SIAM Journal on Applied Mathematics* 51 (October): 1397–1417.

McElroy, William D., and Howard H. Seliger. 1962. Biological luminescence. *Scientific American* (December): 76–89.

Mirollo, Renato E., and Steven H. Strogatz. 1990. Synchronization of pulse-coupled biological oscillators. *SIAM Journal on Applied Mathematics* 50 (December): 1645–1662.

Pennisi, Elizabeth. 1991. Robots go buggy. *Science News* 140 (30 November): 361–363.

Peterson, Ivars. 1996. Keeping the beat. *Science News* 149 (13 April): 236–237.

——. 1996. Step in time. *Science News* 140 (31 August): 136–137.

Simmers, John, Pierre Meyrand, and Maurice Moulins. 1995. Dynamic networks of neurons. *American Scientist* 83 (May/June): 262–268.

Somers, David, and Nancy Kopell. 1995. Waves and synchrony in networks of oscillators of relaxation and non-relaxation type. *Physica D* 89: 169–183.

Stewart, Ian. 1991. All together now. . . . *Nature* 350 (18 April): 557.

Strogatz, Steven H. 1994. *Nonlinear Dynamics and Chaos.* Reading, Mass.: Addison-Wesley.

——. 1994. Norbert Wiener's brain waves. In *Frontiers in Mathematical Biology,* ed. S. Levin. New York: Springer-Verlag.

Strogatz, Steven H., and Ian Stewart. 1993. Coupled oscillators and biological synchronization. *Scientific American* (December): 102–109.

Sullivan, Walter. 1991. A mystery of nature: Mangroves full of fireflies blinking in unison. *New York Times* (13 August).

Terman, D., A. Bose, and N. Kopell. 1996. Functional reorganization in thalamocortical networks: Transition between spindling and delta sleep rhythms. *Proceedings of the National Academy of Sciences USA* 93 (December): 15417–15422.

Williams, Thelma L. 1992. Phase coupling by synaptic spread in chains of coupled neuronal oscillators. *Science* 258 (23 October): 662–665.

Winfree, Arthur T. 1987. *The Timing of Biological Clocks.* New York: Scientific American Library.

——. 1980. *The Geometry of Biological Time.* New York: Springer-Verlag.

5. Different Drums

Cipra, Barry. 1992. You can't hear the shape of a drum. *Science* 255 (27 March): 1642–1643.

Chapman, S. J. 1995. Drums that sound the same. *American Mathematical Monthly* (February): 124–138.

Driscoll, Tobin A. 1997. Eigenmodes of isospectral drums. *SIAM Review* 39 (March): 1–17.

Gordon, Carolyn S. 1989. When you can't hear the shape of a manifold. *Mathematical Intelligencer* 11 (3): 39–47.

Gordon, Carolyn, and David Webb. 1996. You can't hear the shape of a drum. *American Scientist* 84 (January/February): 46–55.

Gordon, Carolyn, David L. Webb, and Scott Wolpert. 1992. One cannot hear the shape of a drum. *Bulletin of the American Mathematical Society* 27 (July): 134–138.

Kac, M. 1985. *Enigmas of Chance: An Autobiography.* New York: Harper & Row.

——. 1966. Can one hear the shape of a drum? *American Mathematical Monthly* 73: 1–23.

Lapidus, Michel L. 1995. Fractals and vibrations: Can you hear the shape of a fractal drum? *Fractals* 3 (4): 725–736.

——. 1991. Fractal drum, inverse spectral problems for elliptic operators and a partial resolution of the Weyl-Berry conjecture. *Transactions of the American Mathematical Society* 325 (June): 465–529.

Lapidus, Michel L., and Michael M. H. Pang. 1995. Eigenfunctions of the Koch snowflake domain. *Communications in Mathematical Physics* 172: 359–376.

Lapidus, Michel L., et al. 1996. Snowflake harmonics and computer graphics: Numerical computation of spectra on fractal drums. *International Journal of Bifurcation and Chaos* 6 (7): 1185–1210.

Lauterborn, Werner, and Joachim Holzfuss. 1991. Acoustic chaos. *International Journal of Bifurcation and Chaos* 1: 13–26.

Peterson, Ivars. 1994. Beating a fractal drum. *Science News* 146 (17 September): 184–185.

Rossing, Thomas D. 1992. Acoustics of drums. *Physics Today* (March): 40–47.

——. 1982. The physics of kettledrums. *Scientific American* (November): 172–178.

Sapoval, B., Th. Gobron, and A. Margolina. 1991. Vibrations of fractal drums. *Physical Review Letters* 67 (18 November): 2974–2977.

Schafer, R. Murray. 1977. *The Tuning of the World.* Toronto: McClelland & Stewart.

Sridhar, S., and A. Kudrolli. 1994. Experiments on not "hearing the shape" of drums. *Physical Review Letters* 72 (4 April): 2175–2178.

Stewart, Ian. 1992. Beating out the shape of a drum. *New Scientist* (13 June): 26–30.

——. 1988. The beat of a fractal drum. *Nature* 333 (19 May): 206–207.

Weidenmüller, Hans. 1994. Why different drums can sound the same. *Physics World* (August): 22–23.

6. Noise Police

Barg, Alexander. 1993. At the dawn of the theory of codes. *Mathematical Intelligencer* 15 (1): 20–26.

Calderbank, A. R., et al. 1993. A linear construction for certain Kerdock and Preparata codes. *Bulletin of the American Mathematical Society* 29 (October): 218–222.

Cartwright, Mark. 1986. Crackly phones and the schoolgirl problem. *New Scientist* (3 July): 36–40.

Cipra, Barry A. 1994. Coding theorists wring linearity out of nonlinear codes. *SIAM News* (February).

——. 1993. Nonlinear codes straighten up—and get to work. *Science* 262 (29 October): 658–659.

——. 1993. The ubiquitous Reed-Solomon codes. *SIAM News* (January).

Conway, J. H., and N. J. A. Sloane. 1994. Quaternary constructions for the binary single-error-correcting codes of Julin, Best and others. *Designs, Codes and Cryptography* 4: 31–42.

Gallian, Joseph. 1995. Math on money. *Math Horizons* (November): 10–11.

——. 1993. How computers can read and correct ID numbers. *Math Horizons* (winter): 14–15.

——. 1991. Assigning driver's license numbers. *Mathematics Magazine* 64 (February): 13–22.

Hammons, A. Roger, Jr., et al. 1994. The Z4-linearity of Kerdock, Preparata, Goethals, and related codes. *IEEE Transactions on Information Theory* 40 (March): 301–319.

McEliece, Robert J. 1989. Safety in numbers: Protecting data mathemagically. *Engineering and Science* (summer): 27–36.

Perry, Tekla S. 1993. Richard W. Hamming. *IEEE Spectrum* (May): 80–82.

Peterson, Ivars. 1995. The Codemart catalog. *Science News* 147 (4 March): 140–142.

——. 1994. Policing digits. *Science News* 145 (12 March): 170–171.

Ross, Robert N. 1985. Of bits and bytes: How computers recognize mistakes. *Bostonia* (February/March): 24–28.

Saff, E. B., and A. B. J. Kuijlaars. 1997. Distributing many points on a sphere. *Mathematical Intelligencer* 19 (1): 5–11.

Thompson, Thomas M. 1983. *From Error-Correcting Codes through Sphere Packings to Simple Groups*. Washington, D.C.: Mathematical Association of America.

7. Complete Chaos

Amato, Ivan. 1992. Chaos breaks out at NIH, but order may come of it. *Science* 256 (26 June): 1763–1764.

Broomhead, Dave. 1990. Keeping a close check on chaos. *Physics World* (July): 23.

Browne, Malcolm W. 1989. In heartbeat, predictability is worse than chaos. *New York Times* (17 January).

Ditto, William L., and Louis M. Pecora. 1993. Mastering chaos. *Scientific American* (August): 78–84.

Ditto, W. L., S. N. Rauseo, and M. L. Spano. 1990. Experimental control of chaos. *Physical Review Letters* 65 (24 December): 3211–3214.

Eckmann, Jean-Pierre, and David Ruelle. 1995. Chaos on demand. *Physics World* (October): 25–26.

Garfinkel, Alan, et al. 1992. Controlling cardiac chaos. *Science* 257 (28 August): 1230–1235.

Glass, Leon. 1996. Dynamics of cardiac arrhythmias. *Physics Today* (August): 40–45.

Heagy, James F., Thomas L. Carroll, and Louis M. Pecora. 1994. Experimental and numerical evidence for riddled basins in coupled chaotic systems. *Physical Review Letters* 73 (26 December): 3529.

Kaplan, Daniel T., and Leon Glass. 1992. Direct test for determinism in a time series. *Physical Review Letters* 68 (27 January): 427–430.

Kautz, R. L., and Bret M. Huggard. 1994. Chaos at the amusement park: Dynamics of the Tilt-A-Whirl. *American Journal of Physics* 62 (January): 59–66.

Kiernan, Vincent. 1994. All the fun of chaos in action. *New Scientist* (4 June): 10.

Kostelich, Eric. 1995. Symphony in chaos. *New Scientist* (8 April): 36–39.

——. 1993. Uncertain determinism. *Nature* 365 (9 September): 106–107.

Langreth, Robert. 1991. Engineering dogma gives way to chaos. *Science* 252 (10 May): 776–777.

Lorenz, Edward N. 1993. *The Essence of Chaos.* Seattle: University of Washington Press.

Moon, Francis C. 1992. Coming to grips with chaos. *Nature* (20 February): 675–676.

Ott, Edward, and Mark Spano. 1995. Controlling chaos. *Physics Today* (May): 34–40.

Ott, Edward, et al. 1993. Scaling behavior of chaotic systems with riddled basins. *Physical Review Letters* 71 (20 December): 4134–4137.

Ott, Edward, Celso Grebogi, and James A. Yorke. 1990. Controlling chaos. *Physical Review Letters* 64 (12 March): 1196–1199.

Pease, Roland. 1995. Why chaos is best for hearts and minds. *New Scientist* (17 June): 18.

Peng, C.-K., et al. 1993. Long-range anticorrelations and non-Gaussian behavior of the heartbeat. *Physical Review Letters* 70 (1 March): 1343–1346.

Peterson, Ivars. 1994. Chaos for fun and profit. *Science News* 145 (26 February): 143.

——. 1993. Finding riddles of physical uncertainty. *Science News* 144 (18 September): 180.

——. 1992. Basins of froth. *Science News* 142 (14 November): 329–330.

——. 1991. Achieving control of chaos at high speeds. *Science News* 140 (12 October): 229.

——. 1991. Ribbon of chaos. *Science News* 139 (26 January): 60–61.

Pincus, Steven M. 1994. Greater signal regularity may indicate increased system isolation. *Mathematical Biosciences* 122: 161–181.

Pincus, Steve, and Burton H. Singer. 1996. Randomness and degrees of irregularity. *Proceedings of the National Academy of Sciences USA* 93 (March): 2083–2088.

Ruelle, David. 1994. Where can one hope to profitably apply the ideas of chaos? *Physics Today* (July): 24–30.

——. 1991. *Chance and Chaos.* Princeton, N.J.: Princeton University Press.

Shinbrot, Troy, et al. 1993. Using small perturbations to control chaos. *Nature* 363 (3 June): 411–417.

——. 1992. Chaos in a double pendulum. *American Journal of Physics* 60 (June): 491–499.

——. 1990. Using chaos to direct trajectories to targets. *Physical Review Letters* 65 (24 December): 3215–3218.

Sommerer, John C., and Edward Ott. 1993. A physical system with qualitatively uncertain dynamics. *Nature* 365 (9 September): 138–140.

Yam, Philip. 1994. Chaotic chaos. *Scientific American* (March): 16.

8. Trails of a Wanderer

Asimov, Daniel. 1995. There's no space like home. *The Sciences* (September/October).

Axel, Richard. 1995. The molecular logic of smell. *Scientific American* (October): 154–159.

Bernstein, Peter L. 1996. *Against the Gods: The Remarkable Story of Risk.* New York: Wiley.

Dimotakis, Paul E. 1996. Turbulence, fractals, and CCDs. *Engineering and Science* 59 (3): 22–34.

Ford, Brian J. 1991. Robert Brown, Brownian movement, and teeth marks on the hatbrim. *The Microscope* 39 (3, 4): 161–173.

Hayes, Brian. 1994. Nature's algorithms. *American Scientist* 82 (May/June): 206–210.

Hersh, Reuben, and Richard J. Griego. 1969. Brownian motion and potential theory. *Scientific American* (March): 66–74.

Klafter, Joseph, Michael F. Shlesinger, and Gert Zumofen. 1996. Beyond Brownian motion. *Physics Today* (February): 33–39.

Lavenda, Bernard H. 1985. Brownian motion. *Scientific American* (February): 70–85.

Mantegna, Rosario N., and H. Eugene Stanley. 1995. Scaling behaviour in the dynamics of an economic index. *Nature* 376 (6 July): 46–49.

Peterson, Ivars. 1996. Trails of the wandering albatross. *Science News* 150 (17 August): 104–105.

Pimbley, Joseph M. 1997. Physicists in finance. *Physics Today* (January): 42–46.

Price, John F. 1996. Optional mathematics is not optional. *Notices of the American Mathematical Society* 43 (September): 964–971.

Schroeder, Manfred. 1991. *Fractals, Chaos, Power Laws: Minutes from an Infinite Paradise.* New York: W. H. Freeman.

Shlesinger, Michael F. 1989. Levy flights: Variations on a theme. *Physica D* 38: 304–309.

Slade, Gordon. 1996. Random walks. *American Scientist* 84 (March/April): 146–153.

Stanley, Michael H. R., et al. 1996. Scaling behaviour in the growth of companies. *Nature* 379 (29 February): 804–806.

Viswanathan, G. M., et al. 1996. Lévy flight search patterns of wandering albatrosses. *Nature* 381 (30 May): 413–415.

Worsley, Keith J. 1996. The geometry of random images. *Chance* 9 (1): 27–40.

9. Gambling with Numbers

Aguayo, Ricardo, Geoff Simms, and P. B. Siegel. 1996. Throwing nature's dice. *American Journal of Physics* 64 (June): 752–758.

Bassein, Susan. 1996. A sampler of randomness. *American Mathematical Monthly* (June/July): 483–490.

Bown, William. 1993. Gambling on the wrong numbers from Monte Carlo. *New Scientist* (24 April): 16.

Brown, Malcolm W. 1993. Coin-tossing computers found to show subtle bias. *New York Times* (12 January).

Cipra, Barry A. 1994. AAAS '94: Random numbers, art and math. *SIAM News* (August/September).

Compagner, Aaldert. 1991. Definitions of randomness. *American Journal of Physics* 59 (August): 700–705.

Doll, J. D., and D. L. Freeman. 1986. Randomly exact methods. *Science* 234 (12 December): 1356–1360.

Eckhardt, Roger. 1987. Stan Ulam, John von Neumann, and the Monte Carlo method. *Los Alamos Science* 15: 131–137.

Erber, T. 1995. Testing the randomness of quantum mechanics: Nature's ultimate cryptogram? *Annals of the New York Academy of Sciences* 775: 748–757.

Erber, T., and S. Putterman. 1985. Randomness in quantum mechanics—Nature's ultimate cryptogram? *Nature* 318 (7 November): 41.

Ferrenberg, Alan M., D. P. Landau, and Y. Joanna Wong. 1992. Monte Carlo simulations: Hidden errors from "good" random number generators. *Physical Review Letters* 69 (7 December): 3382–3384.

Fisher, Daniel S., Geoffrey M. Grinstein, and Anil Khurana. 1988. Theory of random magnets. *Physics Today* (December): 56.

Hayes, Brian. 1993. The wheel of fortune. *American Scientist* 81 (March/April): 114–118.

Heidel, Raymond. 1996. Slot machines. *Scientific American* (August): 112.

James, F. 1993. Good generators give wrong answers. *Europhysics News* 24: 24.

Knuth, Donald E. 1969. *The Art of Computer Programming*. Reading, Mass.: Addison-Wesley.

Kuusela, T. 1993. Random number generation using a chaotic circuit. *Journal of Nonlinear Science* 3: 445–458.

Maddox, John. 1994. The poor quality of random numbers. *Nature* 372 (1 December): 403.

——. 1985. The virtues of single atoms. *Nature* 318 (November 7): 11.

Marsaglia, George. 1968. Random numbers fall mainly in the planes. *Proceedings of the National Academy of Sciences USA* 61 (September): 25–28.

Marsaglia, George, and Arif Zaman. 1994. Some portable very-long-period random number generators. *Computers in Physics* 8 (January/February): 117–121.

Matthews, Robert. 1995. It's a lottery. *New Scientist* (22 July): 38–42.

Metropolis, N. 1987. The beginning of the Monte Carlo method. *Los Alamos Science* 15: 125–130.

Park, Stephen K., Keith W. Miller, and Paul K. Stockmeyer. 1993. Remarks on choosing and implementing random number generators. *Communications of the* ACM 36 (July): 105–110.

Peterson, Ivars. 1992. Monte Carlo physics: A cautionary lesson. *Science News* 142 (19 & 26 December): 422.

——. 1991. Numbers at random. *Science News* 140 (9 November): 300–301.

——. 1990. *Islands of Truth: A Mathematical Mystery Cruise.* New York: W. H. Freeman.

Pickover, Clifford A. 1995. *Keys to Infinity.* New York: Wiley.

——. 1995. Random number generators: Pretty good ones are easy to find. *Visual Computer* 11: 369–377.

——. 1991. Picturing randomness on a graphics supercomputer. *IBM Journal of Research and Development* 35 (January/March): 227–230.

Ross, D. E. 1995. Random number generators are chaotic. *Communications of the* ACM 38 (January): 121–124.

Vattulainen, I., and T. Ala-Nissila. 1995. Mission impossible: Find a random pseudorandom number generator. *Computers in Physics* 9 (September/October): 500–504.

Vattulainen, I., T. Ala-Nissila, and K. Kankaala. 1994. Physical tests for random numbers in simulations. *Physical Review Letters* 73 (7 November): 2513–2516.

Wallace, W. E. 1981. Eminent domains. *The Sciences* (May/June): 10–13.

Warnock, Tony. 1987. Random-number generators. *Los Alamos Science* 15: 137–141.

Wolff, Ulli. 1989. Collective Monte Carlo updating for spin systems. *Physical Review Letters* 62 (23 January): 361–364.

10. Lifetimes of Chance

Albers, Don. 1995. Professor of (magic) mathematics. *Math Horizons* (February): 11–15.

Anderson, Ken. 1995. *Coincidences: Chance or Fate?* London: Blandford.

Chaitin, Gregory J. 1994. Randomness and complexity in pure mathematics. *International Journal of Bifurcation and Chaos* 4: 3–15.

——. 1990. A random walk in arithmetic. *New Scientist* 125 (24 March): 44–46.

——. 1988. Randomness in arithmetic. *Scientific American* 259 (July): 80–85.

——. 1975. Randomness and mathematical proof. *Scientific American* 232 (May): 47–52.

Compagner, Aaldert. 1991. Definitions of randomness. *American Journal of Physics* 59 (August): 700–705.

Davies, Paul. 1992. Is nature mathematical? *New Scientist* (March 21): 25–27.

De Duve, Christian. 1996. The constraints of chance. *Scientific American* (January): 112.

Delahaye, Jean-Paul. 1989. Chaitin's equation: An extension of Godel's theorem. *Notices of the American Mathematical Society* 36 (October): 984–987.

Diaconis, Persi, and Frederick Mosteller. 1989. Methods for studying coincidences. *Journal of the American Statistical Association* 84 (December): 853–861.

Efron, Bradley, and Robert Tibshirani. 1991. Statistical data analysis in the computer age. *Science* 253 (26 July): 390–395.

Ford, Joseph. 1983. How random is a coin toss? *Physics Today* (April): 40–47.

Ford, Joseph, and Giorgio Mantica. 1992. Does quantum mechanics obey the correspondence principle? Is it complete? *American Journal of Physics* 6 (December): 1086–1098.

Hanley, James A. 1992. Jumping to coincidences: Defying odds in the realm of the preposterous. *American Statistician* 46 (August): 197–204.

Hively, Will. 1996. The mathematics of making up your mind. *Discover* (May): 90–97.

Johnson, George. 1995. *Fire in the Mind: Science, Faith, and the Search for Order.* New York: Alfred A. Knopf.

Kiernan, Vincent. 1996. There's one born every minute. . . . *New Scientist* (24 February): 14.

Kolata, Gina. 1990. 1-in-a-trillion coincidence, you say? Not really, experts find. *New York Times* (27 February).

——. 1988. Theorist applies computer power to uncertainty in statistics. *New York Times* (8 November).

——. 1986. What does it mean to be random? *Science* 231 (7 March): 1068–1070.

——. 1984. The art of learning from experience. *Science* 225 (13 July): 156–158.

McKean, Kevin. 1987. The orderly pursuit of pure disorder. *Discover* (January): 72–81.

Nunnikhoven, Thomas S. 1992. A birthday problem solution for nonuniform birth frequencies. *American Statistician* 46 (November): 270–274.

Paulson, Richard A. 1992. Using lottery games to illustrate statistical concepts and abuses. *American Statistician* 46 (August): 202–204.

Peterson, Ivars. 1991. Pick a sample. *Science News* 140 (27 July): 56–58.

Pincus, Steve, and Burton H. Singer. 1996. Randomness and degrees of irregularity. *Proceedings of the National Academy of Sciences USA* 93 (March): 2083–2088.

Rao, C. Radhakrishna. 1966. Uncertainty, statistics, and creation of new knowledge. *Chance* 9 (4): 5–11.

Rescher, Nicholas. 1995. *Luck: The Brilliant Randomness of Everyday Life.* New York: Farrar, Straus & Giroux.

Simon, Julian L. 1994. What some puzzling problems teach about the theory of simulation and the use of resampling. *American Statistician* 48 (November): 290–293.

——. 1987. 'Batter's slump' and other illusions. *Washington Post* (9 August).

Simon, Julian L., and Peter Bruce. 1991. Resampling: A tool for everyday statistical work. *Chance* 4 (1): 22–32.

Steen, Lynn A. 1975. Order from chaos. *Science News* 107 (3 May): 292–293.

Stewart, Ian. 1996. Think maths. *New Scientist* (30 November): 38–42.

Tversky, Amos, and Thomas Gilovich. 1989. The cold facts about the "hot hand" in basketball. *Chance* 2 (1): 16–21.

Wardrop, Robert L. 1995. Simpson's paradox and the hot hand in basketball. *American Statistician* 49 (February): 24–28.

Weaver, Warren. 1963. *Lady Luck: The Theory of Probability.* Garden City, N.Y.: Doubleday.

Acknowledgments

This book represents a distillation of more than fifteen years of reporting and writing for *Science News*. During those years, I have had the opportunity to delve into innumerable topics, covering a wide range of disciplines. In pursuing these ventures into how things work, I have always had a strong appreciation of the crucial role that mathematics plays in helping us make sense of the natural world.

I have also been fascinated by the limits of human knowledge. Physical theory and experiment provide laws and constants to define the world in which we live. Mathematical logic provides clear answers to precisely defined questions. Yet, in all these efforts to grasp the meaning of what we see around us, we constantly face the random and the unknowable. As in the Heisenberg uncertainty principle of quantum mechanics, we can sometimes define with remarkable precision the extent and shape of our ignorance. But nothing rids us of the element of chance.

In our lives, we deliberately carve out pockets of reason and logic; we endeavor to circumscribe and control chaos and chance. Out of these hard-won certainties, we build our remarkable technologies, and in our everyday lives we find ways to negotiate the jungles of uncertainty that constantly press in upon us. So, we continue to learn, calculate, organize, classify, and predict; and we make great strides.

Science involves finding formulas—the physical laws that we believe underlie the machinations of nature. Mathematics tells us that given a formula, it's relatively easy to compute answers. The task is far more difficult, however, when we are given the answers and we must work out the formulas. We collect data—strings of numbers—and from patterns among the numbers in those strings we make stabs at the formulas. As the strings get longer and the numbers become more precise, the formulas get better, and they can work amazingly well. Yet we have no guarantee that we have the right formulas, that we are applying them appropriately, or even that the formulas exist at all. We could be trapped in a local, accidental pattern in an otherwise random existence. All we can do is keep testing our hypotheses and revel in the curious freedom from the straitjacket of determinism that chance and inevitable uncertainty provide.

I am grateful to my editors at *Science News*, first Joel Greenberg, then Patrick Young, and now Julie Miller, for allowing me the freedom to explore and wander the tortuous paths of scientific and mathematical research. Many of the topics covered in *The Jungles of Randomness* have been the subject of articles in *Science News*. Indeed, portions of the material in this book previously appeared, in a somewhat different form, in the magazine.

Undertaking a book of this scope would not have been possible without the cooperation of a large number of people who freely shared with me their ideas, experiences, and insights. I am particularly indebted to Ron Graham, who reviewed the entire manuscript and, along the way, patiently explained mathematical concepts to me and provided illuminating examples and delightful anecdotes that illustrated many aspects of mathematical life.

I also greatly appreciate the help of James Alexander, Bonnie Berger, Bill Ditto, Joseph Gallian, Richard Guy, Nancy Kopell, Michel Lapidus, Cliff Pickover, Michael Shlesinger, Gene Stanley, Steve Strogatz, and Jean Taylor, who contributed in a variety of ways. My apologies to anyone I have inadvertently omitted from the list.

My thanks go to Robert Dickau, formerly of Wolfram Research, who provided wonderful advice on how to create several illustrations for the book using *Mathematica 3.0*. Gandhi Viswanathan of Boston University went to a great deal of extra trouble to produce beautiful visualizations of random walks especially for the book.

My wife, Nancy, played a major role in editing the original manuscript. She and our two sons, Eric and Kenneth, contributed in a number of ways to the making of this book. I also wish to thank Gail Ross, my agent, and Emily Loose, my editor at Wiley, for helping bring this book into existence within a reasonable time—certainly more quickly and efficiently than a host of monkeys banging randomly on typewriter keys would have managed to do.

Index